Springer Climate

Series Editor

John Dodson , Institute of Earth Environment, Chinese Academy of Sciences, Xian, Shaanxi, China

Springer Climate is an interdisciplinary book series dedicated to climate research. This includes climatology, climate change impacts, climate change management, climate change policy, regional climate studies, climate monitoring and modeling, palaeoclimatology etc. The series publishes high quality research for scientists, researchers, students and policy makers. An author/editor questionnaire, instructions for authors and a book proposal form can be obtained from the Publishing Editor.

Now indexed in Scopus® !

More information about this series at http://www.springer.com/series/11741

Md. Jakariya · Md. Nazrul Islam
Editors

Climate Change in Bangladesh

A Cross-Disciplinary Framework

 Springer

Editors
Md. Jakariya
Environmental Sciences and Management
North South University
Dhaka, Bangladesh

Md. Nazrul Islam
Geography and Environment
Jahangirnagar University
Dhaka, Bangladesh

ISSN 2352-0698   ISSN 2352-0701   (electronic)
Springer Climate
ISBN 978-3-030-75827-1   ISBN 978-3-030-75825-7   (eBook)
https://doi.org/10.1007/978-3-030-75825-7

This Springer imprint is published by the registered company Springer Nature Switzerland AG
The registered company address is: Gewerbestrasse 11, 6330 Cham, Switzerland

Preface

Climate change and its variability pose serious challenges in today's global context. Meeting the needs of the most vulnerable through adequate resources is a huge concern nowadays. However, climate finance debate being present in the center stage of global negotiations for decades only deepens its importance as a global issue. Along with the inherent difficulty to address it because of a lack of a proper definition, climate finance debate has taken its turns through various challenging discourses. In terms of institutional framework, together with the Global Environment Facility (GEF), the newly established Green Climate Fund (GCF) will serve as the financial mechanism of the Convention. Also, some other funds established under the decisions of COP7 in 2001 and operationalized few years later will continue serving the climate regime. Many adaptation projects have the typical characteristics of development projects, which makes it difficult to convince the donors in approving such projects. However, it is important to understand that in certain countries, vulnerability of a society is dependent on the structural development, and hence, it is not always logical to distinguish adaptation and development. This is because in these situations, development deficits put the vulnerable communities to further risks. The international treaty obligation for the developed countries to support the developing and vulnerable countries is not only about legal binding but also a matter of upholding human rights. For this, democratization of climate finance governance is imperative with core principles to be ensured through the system.

Subsequently, the main environmental threat from biodegradable waste is the production of methane. Biodegradable waste, when collected and processed in an industrial digester, can produce natural gas, used for homes, as well as a growing number of truck and bus fleets in developed nations. Compare this with natural gas, which contains 80 to 90% methane. The energy content of the gas depends mainly on its methane content. High methane content is therefore desirable. A certain carbon dioxide and water vapor content is unavoidable, but sulfur content must be minimized particularly for use in engines. Climate change coupled with anthropogenic disturbances poses a great threat to the existence of this mangrove. Many regions of the world are affected by climate change, but Sundarbans is one of the highest affected regions due to high level of salinity, sedimentation, and land erosion. The salinity is increasing day by day due to frequent cyclones, sedimentation, and brackish tiger prawn cultivation. The increased salinity is jeopardizing the ecosystems of Sundarbans and poses more risk than any other stressors. The study aims to assess the impact of salinity on the pioneer and indicator plant species in terms of species distribution and the coping capacity with the increased salinity. Primary data was collected from 30 sample plots which were fresh swamp and fresh-brackish swamp in the past. Secondary historical data was collected from the Forest Department to understand the natural dynamics.

In Bangladesh, food security has been one of the major national priorities for last few decades but the target has always been interrupted by the climate change and for resource constraints. Present section of this chapter will highlight the major effects of climate change in the food production and the national resources constraints to address the food security. However, major constraints in terms of food security in Bangladesh attributed to cultivable land scarcity, irrigation water scarcity in summer, lack of technological knowledge, lack of climate adaptive crop variety, lack of institutions, and professionals as well as social and cultural constraints. Richer farmers can afford modern machineries, genetically modified crop seeds and chemical fertilizers. This results in efficient farming, higher yield from a unit plot of land, or better utilization of larger farmlands. Climate changes are expecting to contribute to some air quality problems. Respiratory diseases may be exacerbated by warming-induced increased frequency of events and allergen in air. It has been said that ground-level ozone can damage lung tissue, and it is harmful for those who have asthma and other chronic lung diseases. The preparation of Bangladesh to face the challenge of global warming is not enough and cannot be overlooked. In order to tackle the health and socio-economic effects, relevant stakeholders including policy makers, program designers, program implementers, civil servants, and civil society members need to have better understanding about both climate change and its possible impacts. Due to favorable climatic condition, tea industry of Bangladesh is one of the most important sources of income. Sylhet, the northeastern divisional city of Bangladesh, is the major tea-producing region of the country. For this reason, the study area was selected in Sylhet region to find out the causes of fluctuation of recent tea production in the study area.

The analysis depicted that the households not having the access to marketing information were 0.214 times significantly less like to be food secure as compared to the households having the same access to the information related to market price of input, output, and materials needed for shrimp culture. Provision of training, in this study, was significantly associated with the food security status. Expansion of metropolitan area poses a greater risk toward the environment which needs immediate attention to the problem of solid waste disposal, air pollution control, and deterioration of the urban environment. The average highest generation rate was found to be 0.368 kg/capita/day at residential areas in Dhaka, whereas the lowest was 0.259 kg/capita/day in Barisal has been discussed in this book. Climate change is predicted to impact on fisheries and dependent communities. This study assesses the vulnerability and adaptation to the impacts of climate variability and change, in three small-scale coastal fishing communities in Bangladesh with a view to suggest policy and scaling-up the findings. Overall, using a mixed method approach, this study contributes empirical evidence to current debates in the literature on climate change by enhancing an understanding of the characteristics and determinants of livelihood vulnerability, migration as an adaptation strategy, and limits and barriers to the adaptation of fishing communities to climate variability and change.

Due to climatic Change in the recent years, the existing national database of Bangladesh lacks information on lightning casualties. Hence, a five years of database on lightning related deaths and injuries from 2011–2016 were constructed through an innovative data mining process. An average of 913 casualties was identified, with an average of 182 people being affected by lightning occurrences each year in Bangladesh. The largest death toll was found among the male population (74%) compared to the females (26%), as males are more involved with labor-intensive agricultural practices in a developing country like Bangladesh. In Bangladesh, fisheries contribute about nearly 3% to GDP and more than 8% to the export earnings of the country (Bangladesh Population and Housing Census 2011, 2015). Marine fish contributes about 20% of total fish production in Bangladesh. (Islam et al. 2001). Marine fisheries constitute of industrial fishery by large trawlers and artisanal fisheries by mechanized and non-mechanized boats.

Dhaka, Bangladesh   Professor Dr. Md. Jakariya
                                                             Professor Dr. Md. Nazrul Islam

Contents

1 **Climate Finance in the UNFCCC Negotiations: Bridging Gaps
   with Lessons Learnt**   1
   Sirazoom Munira, Raisa Bashar, Tahmid Huq Easher,
   and Mizan R. Khan

2 **Climate Change and State of Renewable Energy in Bangladesh:
   An Environmental Analysis**   25
   Kamrun Nahar and Sanwar A. Sunny

3 **Climate Change Impact on Sundarbans: Challenges
   for Mitigation Strategies**   47
   Md. Mizanur Rahman, Md. Rakib Hossain, and Md. Nazrul Islam

4 **Climate Change and Sustainability of Agriculture
   in Bangladesh**   65
   Nazmul Ahsan Khan

5 **Climate Change and Its Impact on Health in Bangladesh**   85
   Mohammad Delwer Hossain Hawlader

6 **Climatic and Environmental Challenges of Tea Cultivation
   at Sylhet Area in Bangladesh**   93
   Md. Nazrul Islam, Sahanaj Tamanna, Md. Mizanur Rahman,
   Mohammad Ahmmed Ali, and Imran Mia

7 **Determinants of Food Security in the Environmentally Stressed
   Areas in Bangladesh**   119
   Mohammad Amirul Islam, Avik Chowdhury, Md. Anisuzzaman,
   Md. Shohanur Rahaman Shetu, Khandaker Md. Mostafizur Rahman,
   and Moupia Rahman

8 **Climate Change and Municipal Solid Waste Management
   in Dhaka Megacity in Bangladesh**   135
   Hassan Mahmud

9   **Impacts and Responses of Bangladesh Coastal Fishing
     Communities to Climate Change: Implications for Policy
     and Scaling-up** 157
     Md. Monirul Islam

10   **Climate Change and Lightning Risk in Bangladesh** 183
     Fahmida Kabir and Md. Jakariya

11   **Climate Change Impact and the Conservation of Marine Turtles:
     A Case Study from Teknaf, Bangladesh** 205
     Methila Sarker, Alifa Bintha Haque, Md. Nazrul Islam,
     and Md. Jakariya

About the Editors

Prof. Md. Jakariya specializes in Human Geography, Environmental Risk Assessment, Climate Change Adaptation, Water Resources Management and GIS. Prof. Jakariya is currently working as Professor of the Department of Environmental Science and Management of North South University (www.northsouth.edu). He has a wide range of experience in designing and implementing action research in the field of environment and development. He worked as a lead researcher for many national and international organizations mostly on natural resources management and sustainable development issues for more than two decades. Under his leadership, the department has seen notable success including the launch of two environmental and GIS labs. In addition to leading the department and teaching, Prof. Jakariya is an Executive Member of the National Environmental Committee. Before joining NSU, he was one of the pioneering contributors to the sustainable arsenic mitigation projects of UNICEF, Bangladesh. His contribution to providing sustainable safe water techniques for underprivileged people of Bangladesh is praiseworthy. He is currently the lead investigator of several projects funded by national and international donors. He received MPhil degree in Environment and Development from University of Cambridge, UK, in 2000, and Ph.D. from the Royal Institute of Technology (KTH), Sweden in 2007. Prof. Jakariya has published a good number of papers in international peer-reviewed journals and books.

Professor Dr. Md. Nazrul Islam is a permanent Professor at the Department of Geography and Environment in Jahangirnagar University, Savar, Dhaka-1342, Bangladesh. Prof. Nazrul has completed his Ph.D. from the University of Tokyo, Japan. Besides, he has completed Two Year Standard JSPS Postdoctoral Research Fellow from the University of Tokyo, Japan. Prof. Nazrul fields of interest are environmental and ecological modeling, climate change impact on aquatic ecosystem, modeling of phytoplankton transition, harmful algae, and marine ecosystems regarding to deal with hydrodynamic ecosystems coupled model on coastal seas, bays and estuaries, application of computer based programming for numerical simulation modeling etc. Prof. Nazrul is an expert on scientific research techniques and methods to develop the models for environmental systems analysis research. Prof. Nazrul also visited as an invited speakers in several foreign Universities in Japan, USA, Australia, UK, Canada, China, South Korea, Germany, France, the Netherlands, Taiwan, Malaysia, Singapore and Vietnam etc. Prof. Nazrul has been awarded "Best Young Researcher Award" by the International Society of Ecological Modeling (ISEM) for the outstanding contribution to the Ecological Modeling fields, 2013, Toulouse, France and he has also been awarded "Best Paper Presenter Award 2010" by SautaiN in Kyoto, Japan. He has also been awarded as a "Best Poster Presenter Award" in the Techno Ocean Conference, Kobe City Exhibition Hall at Kobe in Japan.

Prof. Nazrul has made more than 40 scholarly presentations in more than 20 countries around the world, authored more than 120 peer-reviewed articles and authors of 10 books and research volumes. Currently Prof. Nazrul has published an excellent textbook entitled *Environmental Management of Marine Ecosystems* jointly with Prof. Sven Erik Jorgensen by the CRC press (Taylor & Francis). He has also currently published an excellent book entitled: *Bangladesh I: Climate Change Impacts, Mitigation and Adaptation in Developing Countries,* Springer

Publication, the Netherlands and Germany. Prof. Nazrul is currently serving as an *"Executive Editor-in-Chief"* of the journal *Modeling Earth Systems and Environment*, Springer International Publications (Journal no. 40808). E-mail: nazrul_geo@juniv.edu.

Chapter 1
Climate Finance in the UNFCCC Negotiations: Bridging Gaps with Lessons Learnt

Sirazoom Munira, Raisa Bashar, Tahmid Huq Easher, and Mizan R. Khan

Abstract Climate finance debate being present in the centre stage of global negotiations for decades only deepens its importance as a global issue. Along with the inherent difficulty to address it because of a lack of a proper definition, climate finance debate has taken its turns through various challenging discourses. Regardless of these, there have been several incidents of consensus, not only in thinking but in collective action—as demonstrated by many developed and developing country Parties to address challenges but also taking actions. These actions have helped to not only bridge gaps at the global negotiation tables, but to work on the past mistakes and make way for a more transparent and reliable climate finance forum. In addition, debate over adaptation finance and development finance is currently another big issue. This stems from the fact that many adaptation projects have the typical characteristics of development projects, which makes it difficult to convince the donors in approving such projects. However, it is important to understand that in certain countries, vulnerability of a society is dependent on the structural development, and hence, it is not always logical to distinguish adaptation and development. This is because in these situations, development deficits put the vulnerable communities to further risks. The international treaty obligation for the developed countries to support the developing and vulnerable countries is not only about legal binding, but also a matter of upholding human rights. For this, democratization of climate finance governance is imperative with core principles to be ensured through the system. These principles include accountability, transparency along with public and gender-equitable participation in the decision making process. Furthermore, to reach a consensus, an understanding of where the gaps are occurring in opinions between the donors and the recipients is also key to address and then effectively bridge the gaps.

S. Munira (✉) · R. Bashar · T. H. Easher · M. R. Khan
Environmental Science and Management (ESM) Department, North South University,
Bashundhara R/A, Dhaka 1229, Bangladesh
e-mail: sirazoom.munira@northsouth.edu

© Springer Nature Switzerland AG 2021
Md. Jakariya and Md. N. Islam (eds.), *Climate Change in Bangladesh*,
Springer Climate, https://doi.org/10.1007/978-3-030-75825-7_1

Keywords Climate finance · Global issues · UNFCCC · Negotiations · developing country

Introduction

Climate change and its variability pose serious challenges in today's global context. Meeting the needs of the most vulnerable through adequate resources is a huge concern for Parties who make important decisions at the Conference of Parties (COP) under the United Nations Framework Convention on Climate Change (UNFCCC) negotiation process. So far, the Paris Agreement has been one of the major achievements in the climate discourse, whereby UNFCCC achieved a great diplomatic success in bringing all Parties to a common platform and agreeing on major building blocks, which included mitigation, adaptation, loss and damage, finance, technology development and transfer, capacity building among others. The UNFCCC acknowledges the climate change-induced risks and vulnerabilities and calls for special efforts to reduce the impacts. The Articles 3.1, 4.3, 4.4, 4.5 and 4.9 (from the consolidated version of the convention text including amendments to Annex I and Annex II) of the convention mentioned the responsibilities of the developed country Parties towards the developing countries that are particularly vulnerable to the adverse effect of climate change (UNFCCC 1992).

In retrospect, providing finance to the countries which are vulnerable to impacts of climate change was a fundamental part of the UN Rio Treaty under the UNFCCC in 1992. Since then, there has been a number of general agreements which press on the urgent need for funds to be dedicated to climate change activities, which is echoed in the Bali Action Plan (BAP). There has also been focus on the scale of funding, but with little discussion about the sourcing of the additional resource that is required. However, transparency, modality and accounting of these financial resource flow have been an issue of several security in the past.

Against this backdrop, this chapter attempts to summarize the legal and institutional framework on which climate finance (CF) is based under the UNFCCC. It outlines the inherent difficulty caused due to the absence of a universally accepted definition of climate finance and the key debates around the issue and also demonstrates the current state of climate finance in Bangladesh, as well as at a global level. It contains a gist of the strategic actions of Bangladesh outlining its take on climate change as a vulnerable country with very little financial and technical resources to combat its impacts. The section outlining key debates in climate finance summarizes the challenges surrounding this issue down in the grassroots straight up to the Convention. However, climate finance debate being present in the centre stage of global negotiations for decades only deepens its importance as a global issue. Along with the inherent difficulty to address it because of the lack of a proper definition, climate finance debate has taken its turns through various challenging discourses. Regardless of these, there have been several incidents of consensus, not only in thinking but in collective action—as demonstrated by many developed and developing country Parties to address challenges, but also taking actions. These actions have helped to not only bridge gaps at the global negotiation tables, but to work on the past mistakes and make way for a more transparent and reliable climate finance forum.

Additionally, the different definitions of CF proposed by organizations is highlighted in this chapter, along with an analysis of the Subsidiary Body for Scientific and Technological Advice (SBSTA) submissions by the donor, recipient and observer Parties to evaluate the differences and similarities among their views. The authors believe that the differences in views involving capacity building, fund weighing mechanisms, reporting formats and diversification of projects portray the long road ahead of the negotiations. However, the unsaid consensus/common grounds on private and public intervention, roles of MDBs, improvement of reporting formats and monitoring mechanisms demonstrate that there may be a way out of the "blame game." This, in turn, will help identify the starting points to solve the problems of the missing universal climate finance definition and framework, as well as pave the way towards mitigating the transparency and accountability issues.

Current Status of CF: The World Arena

The UNFCCC provided a legal framework and guiding principles that define the climate finance governance in the world. When the UNFCCC was adopted, Parties unanimously realized that climate change calls for the "widest possible cooperation between the countries and their participation in an effective and appropriate international response, in accordance with their common, but differentiated responsibilities and respective capabilities and their social economic conditions" (UNFCCC 1992). The Convention provided legal framework to support the implementation of mitigation and adaptation programmes and projects (Ludemann and Ruppel 2013). This emphasized on the cooperation and contribution of countries involved to assess their specific responsibilities and respective capacities. It said that these state actors ought to consider financial implications for their financial responsibilities. The Paris Agreement also reaffirmed the need for financial assistance and cooperation from the developed countries to address climate change (UNFCCC 2015).

Legal and Institutional Framework of Climate Finance

The Convention lays down the basic principles of economics and financing for addressing climate change in Article 3 which include equity and common, but differentiated responsibility based on respective capabilities (CBDR-RC), consideration of specific needs and special circumstances, especially of those that are particularly vulnerable to climate impacts, ensuring cost-effectiveness and global benefits from adopted measures and recognition of the right to promote development. The commitments under the climate regime obligate developed countries to provide financial support (Article 4.3) for adaptation and mitigation projects in developing countries (Articles 4.1, 4.3, 4.4, 4.7, 12.1). As mentioned previously, Articles 4.3 and 4.4 can be taken as prominent reflections of the CBDR principle. Furthermore, adaptation has been recognized as a global responsibility under the PA, setting a global goal and linking it to the level of mitigation. This is a step forward in elevating the importance and urgency of adaptation in the developing countries, but in terms of finance, not much progress is there yet.

In terms of institutional framework, together with the Global Environment Facility (GEF), the newly established Green Climate Fund (GCF) will serve as the financial mechanism of the Convention. Also, some other funds established under the decisions of COP7 in 2001 and operationalized few years later will continue serving the climate regime. These funds are the LDC Fund, the Special Climate Change Fund and the Adaptation Fund. The GCF has been capitalized with a fund of USD 10.3 billion. However, so far only about 5.6% of total CF has been delivered through these somewhat democratically administered funds. Besides these funds, quite a number of bilateral and multilateral agencies also deliver CF (Oxfam 2012).

The Paris Agreement, which brought Parties to agree to combat climate change and accelerate actions towards a sustainable future, is a landmark agreement bringing all nations to a common cause and realizing these framework(s). According to Articles 9, 19 and 11, it reaffirms the obligations of the developed countries to support the developing country Parties to build a climate-resilient future. Also, calling for voluntary contributions by other Parties, the agreement calls for provision of resources that ought to keep a balance between adaptation and mitigation activities. The Paris Agreement provides that the Financial Mechanism of the Convention, which includes the GCF shall serve the Agreement (What is the Paris Agreement? n.d.). The Paris Agreement also emphasizes on transparency (Article 13) and implementation and compliance (Article 15)—it relied on a robust transparency and accounting system in order to provide clarity on climate actions. With that a "global stocktake" to take place in 2023 and every five years after that the collective progress should be assessed towards achieving the purpose of the Agreement.

Issues related to climate finance have been a subject of debate at global platforms. The debate surrounds ideas regarding how climate finance is sourced and mobilized through a number of financial instruments and channels. It includes the kind of support and the extent to which it can be assumed as "new and additional" in order to support the countries that are really in need. A key area of debate also surrounds the idea of "fair share" of climate finance and how to assess whether this amount of funding is the fair amount (Bird 2014). On the other hand, there has been a lot of attention on mitigation finance in the past, which calls for a balanced support for both adaptation and mitigation activities. Parties to UNFCCC, despite of agreeing to this balanced approach, still have a lack of understanding on how this is to be interpreted in practice (Nakhooda 2013; Nakhooda and Norman 2013). Recent climate negotiations have been focused on strengthening focus on adaptation and increasing financial capacities to address it (Ludemann and Ruppel 2013).

However, unlike mitigation, which can be tracked through the greenhouse gas emission reduction, there has not yet been a metric system to track progress of adaptation. In addition, debate over adaptation finance and development finance is currently another big issue. This stems from the fact that many adaptation projects have the typical characteristics of development projects, which makes it difficult to convince the donors in approving such projects. However, it is important to understand that in certain countries, vulnerability of a society is dependent on the structural development, and hence, it is not always logical to distinguish adaptation and development. This is because in these situations, development deficits put the vulnerable communities to further risks. So, any development project that aims to mitigate structural problems also have a large contribution to increasing the society's adaptation capacity (Weischer and Wetzel 2017).

Also, major climate policy agreements made so far, including that of the Kyoto Protocol and the Cancun Agreement, have mentioned that funds that address climate change for the developing countries must be new and additional. Although these two terms are indicative of being over and above development aid, they have never been defined properly (Stadelmann et al. 2011). It is also important to ensure that achieving major reduction in greenhouse gas emission and building capacity of the vulnerable communities to combat climate change is being done at a very fast rate; however, this must not take place by sacrificing the needed development.

The PVCs having nano-contributions to causing the climate change problem are being hit first and hardest as innocent victims, with extremely weak adaptive capacity on their own (Khan 2013). The developed countries must account for the greenhouse gas emissions which lead to major implications, especially in countries that are geographically vulnerable to extreme events. Hence, the transfer of these resources cannot only be a legally binding regime but also an issue of human rights (Schalatek 2012). Also, a variety of new arrangements must be facilitated to generate public and private climate finance, and a "single uniform design is neither feasible nor desirable". These designs should support and "not retard" the future adoption by many developing nations of emission caps (Stewart et al. 2009).

Parties and experts who aim to mobilize climate finance at scale proposed to have alternate sources of climate finance which includes trading schemes, carbon taxes, etc. However, the United States (US) pulling out of the Paris Agreement made this scenario quite complicated. The US, as one of the world's largest financial superpower, has a strong ability to raise financial and technological resources. Its contribution is key to mobilizing funds and is unmatched, especially to make the latest technology available to the rest of the world, and hence, its withdrawal from the Paris Agreement will have serious implications in the climate finance arena. The Parties to the Paris Agreement, with their commitments to meet resilience targets and also cut down greenhouse gas emissions, have also set some financial targets. Although developing nations made their commitments conditional to receiving global support through international climate funds, there is still not enough attention to assessing the legal frameworks of the developing countries. It is important for the developing countries to identify the legal barriers and opportunities to optimize options for climate finance to fund the Nationally Determined Contributions (NDCs) of these countries (Morita and Pak 2018).

Current Status of CF: Bangladesh

There are many different elements to climate finance. These include the type of finance provided (e.g. development aid, equity, low cost loans, etc.), the sources of this finance (is it public or private?) and where the finance flows from (developed countries to developing countries, within developed nations or from other sources). Financing can be provided through a number of channels including bilateral, regional or other multilateral channels along with envisaged financial mechanism. This has been clarified in paragraph 5 of Article 11 of UNFCCC and paragraph 3 of Article 11 of Kyoto Protocol. Ludemann and Ruppel (2013) echo that this opens many opportunities for state actors and relevant stakeholders to play important roles in financing activities dedicated to climate change.

Climate finance flows went up to USD 437 billion in 2015 until falling 12% in 2016 to USD 383 billion. The surge was due to private renewable investments, in China and in the US. However, the fall in climate finance flow in 2016 was because of lower capacity additions in a number of countries along with falling technology costs (Buchner et al. 2017).

According to the Intergovernmental Panel on Climate Change (IPCC), Bangladesh is one of the most vulnerable countries to climate change. However, the country has taken large strides to combat climate change through long-term strategic approach. Bangladesh has prepared the Climate Fiscal Framework (CFF) by the partnership of the Finance Division of Bangladesh and Poverty Environment and Climate Mainstreaming (PECM) project of the General Economic Division; it has been developed in line with the Bangladesh Strategy and Action Plan (BCCSAP) 2009, the Sixth Five-Year Plan and other important initiatives of the Bangladesh government. The CFF provides such tools by identifying the supply and demand sides of the climate fiscal funds. However, the framework could be reviewed to ensure that transparency, accountability and sustainability aspects in climate finance are existent in the long run. Bangladesh has been setting examples in terms of governance of climate funds too. The government not only implemented the BCCSAP 2009 by the establishment of the Bangladesh Climate Change Trust Fund, but also enacted the Climate Change Trust Act in 2010. Large funds were allocated to BCCTF, supporting a number of projects undertaken and implemented by various ministries, departments, NGOs and others. Strategic Program for Climate Resilience (SPCR) is another financing facility to address climate threats in Bangladesh (Climate Fiscal Framework 2014). A comprehensive climate finance framework is key to achieving climate finance readiness in a nation. For such framework, the inclusion of some key factors is critical. According to a UNDP's discussion paper, Development in a Changing Climate: A framework for Climate Finance in 2010, financial planning refers to one of its key components which does not only assess the needs and priorities of climate-related activities, but also identifies policy mix and source of financing. It also presents the importance of assessing finance through blends and combinations to formulate projects, programmes and sector-wide approaches to access finance.

On the other side, there should be proper channels for delivery of the funds and to coordinate implementation and execution of projects, programmes, sector-wide approaches among others. Lastly, the components should be measureable, reportable and verifiable (MRV), and if performance-based payments are to be used, MRV is key. On that note, it is essential for Bangladesh to be made ready to use the national and international climate finance in the most effective way and this raises the need of a holistic framework. A comprehensive climate finance framework should provide guidelines to track climate-related expenditures and estimate the potential costs that may be required to address future climate change challenges. To formulate climate change policies, a climate finance framework aids to provide essential tools, guidelines and principles to achieve the goals of a climate fiscal policy without intruding into public finances. Integration of a climate finance framework with the governance structure of Bangladesh is important. It should help provide performance management indications at the ministry levels to reinforce accountability. This is also one of the key recommendations of the Climate Public Expenditure and Institutional Review (CPEIR), which states that the budget formulation and its execution must address climate change. Also, the framework should have adequate scope for auditing to be conducted on climate activities and also have close engagement and combined ownership of both the public and the private sectors. Bangladesh, as an LDC, should raise its voice to have climate funds as grants, especially for adaptation, as promised at Copenhagen in 2009 by the developed countries.

It is very important for countries like Bangladesh to be well-equipped to meet the standards set by the Green Climate Fund (GCF) and other adaptation fund windows as there is a race in accessing funds. Enhanced access capacity and effective utilization of the funds are a must, along with strategically using political leverage for raising funds. However, it is imperative for Bangladesh to have improved governance mechanism to have better access to global climate funds. This will be possible through multi-stakeholder engagement and maintenance of transparency and accountability through the whole governance process. For this, Bangladesh has to identify key challenges that exist in their climate finance discourse. The recipient countries also need to strengthen their institutional capacity through improved public–private partnership to have an enabled environment for financing climate change. In Bangladesh, there should be an enabling environment to finance private sector in addressing climate change. Private sector's engagement is insufficient in decision-making or policy implementation, especially in countries like Bangladesh. Their inclination is to engage this sector through public–private partnership through financial incentives. However, it is very important for the vulnerable countries to identify their vulnerable sectors for just climate finance allocation.

Attempts at Defining Climate Finance

Comparative analysis of the various definitions of climate finance which has been proposed by various organizations show a large amount of similarity. Although there are a number of views of what type of funding makes up "climate finance", there is no standard definition of the term (Venugopal and Patel 2013). Referring broadly to financial resources that are dedicated to cover the costs of transitioning to a low-carbon global economy and to address activities to build resilience against the current and future threats of climate change (CPI 2014), climate finance is yet to receive a universally accepted definition. But broadly, at international negotiations in global climate politics regime, the term is used to describe financial flows from the developed to the developing nations to address mitigation and/or adaptation activities (Venugopal and Patel 2013).

Climate finance was first attempted to be defined by the Rio 1992 UNFCCC report (UNFCCC 1992) as the following: developed countries shall provide "new and additional financial resources" to developing countries to support meeting the full and incremental costs of climate change. Hence, there is a strong legal binding for the developed countries to support the developing countries in combating climate change impacts to become more resilient and also less carbon-intensive. The concept of "incrementality" or "additionality" plays an important role in understanding the sum of climate funds, which must be "new and additional" (Brown et al. 2010; Stadelmann et al. 2011). Climate funds should therefore be funds that are mobilized from new sources, such as levy or emissions trading; it should be delivered through new channels which includes climate finance windows like the GCF, and it should be in excess of current climate finance (Brown et al. 2010; Stadelmann et al. 2011).

Although UNFCCC does not have a definition of climate finance, in its report on Standing Committee on Finance on the 2014 Biennial Assessment and Overview of Climate Finance Flows, it has tried to point to a convergence of (operational) definitions that is as follows: *Climate finance aims at reducing emissions, and enhancing sinks of greenhouse gases and aims at reducing vulnerability of, and maintaining and increasing the resilience of, human and ecological systems to negative climate change impacts* (UNFCCC 2014). UNFCCC's report on Standing Committee on Finance on the 2014 Biennial Assessment and Overview of Climate Finance Flows also mentions, "*...the report encountered challenges in collecting, aggregating and analysing information from diverse sources. For example, each of these sources uses its own definition of climate finance and its own systems and methodologies for reporting. The wide range of delivery channels and instruments used for climate finance also poses a challenge in quantifying and assessing finance*".

ADB's Sustainable Development Working Paper Series No. 34 (Chandrasekhar 2014) almost rightly points where the problem is being created as a consensus is yet to be reached on defining "climate finance". The paper refers: "*There is no agreed definition of climate finance. Consequently, accurate data on international climate finance are not available. Data on bilateral and multilateral flows are available, but they involve some double counting. In some cases the data reflect the total financial commitment to a project that has adaptation or mitigation benefits rather than the share of the project cost attributable to the climate objective.* ADB has simply put the definition of climate finance in one of its 2014 report as the *financial support by industrialized countries for adaptation and mitigation actions in developing countries*". On the other hand, the Organization for Economic Co-operation and Development's Development Assistance Committee (OECD-DAC) has no definition on climate finance as well; instead, the OECD-DAC defines and reports on climate-related Official Development Assistance (ODA) and have five (5) statistical markers for monitoring external development markers for environmental purposes (UNFCCC 2014).

The Particularly Vulnerable Countries (PVCs) have negligible contributions to greenhouse gas emissions across the globe. However, they are hit first and hardest, with very weak adaptive capacity to tackle the impacts (Khan 2013). This is where the rationale or legal basis emerges for industrial countries to support the PVCs in addressing these impacts. Since climate finance comes from the notion of responsibility-capability-based mechanism (Article 3.1), it is therefore very different from the voluntary-based development aid. Climate funds are identified as "new and additional" and hence are in excess of the 0.7% of Gross National Income (GNI) contribution that is addressed to ODA. Thus, climate finance ought to be (a) in excess of current ODA, (b) in excess of ODA levels from a specified baseline year and (c) in excess of projected ODA calculated using a specified formula (Brown et al. 2010).

On the contrary, from the Biennial Assessment and Overview of Climate Finance Flows (2014), it is clear that the Multilateral Development Banks (MDBs) understand that climate finance is equal to the sum of mitigation, adaptation and dual benefit finance from the MDB own resources as well as external resources. The Biennial Assessment and Overview of Climate Finance Flows (2014) states that the International Development Finance Club mentioned about "Green finance" which comprises "climate finance" and finance for "other environmental objectives", with "climate finance" being composed of "green energy and mitigation of greenhouse gases" and "adaptation to climate change.

Furthermore, according to the World Resources Institute (WRI), climate finance —or international climate finance—is used to describe financial flows from developed to developing countries for climate change mitigation/adaptation activities. In contrast, Climate Policy Initiative (CPI) puts climate finance in a different angle. They put the idea of climate finance as the financial resources paid to cover the costs of transitioning to a low-carbon global economy and to adapt to, or build resilience against, current and future climate change impacts. On the other hand, the Overseas Development Institute (ODI), which has done a number of works on climate finance instruments and the architecture, refers to climate finance as the financial resources mobilized to help developing countries mitigate and adapt to the impacts of climate change, including public climate finance commitments by developed countries under the UNFCCC. Furthermore, the primary national policy documents by the Government of Bangladesh (GoB) which analysed the concept of climate finance very critically, including the Climate Fiscal Framework (CFF), the Climate Public Expenditures and Institutional Review (CPEIR) and Climate Protection and the Development (CPD) Budget Report 2017–18, could not pinpoint the definition of climate finance as has been mentioned in the Bangladesh Climate Change Strategy and Action Plan (BCCSAP), the Seventh Five-Year Plan (7FYP) and the CPEIR. However, according to Climate Fiscal Framework (CFF) of Bangladesh: the expressions "climate finance", "climate expenditures" and "climate-related expenditures" are used interchangeably in the CFF, which includes both adaptation- and mitigation-related finances and expenditures. This study therefore presses to propose a comprehensive definition of climate finance, which is imperative in understanding what is inclusive and/or is exclusive when climate finance is acknowledged.

Hence, the central idea remains that climate finance aims at reducing emissions, and enhancing sinks of greenhouse gases and aims at reducing vulnerability of, and maintaining and increasing the resilience of, human and ecological systems to negative climate change impacts. This has been the overall understanding obtained from United Nations Framework Convention on Climate Change (UNFCCC).

Issues of Climate Finance

Accountability and Transparency in the Climate Finance Arena

Several meetings on aid effectiveness have taken place from 2005 onwards on the principles of mutual accountability, transparency and shared responsibility, ownership and partnership for CF from the industrial nations to the developing, most vulnerable nations. It was assumed that if the transparency was ensured then accountability would result accordingly and ownership of aided projects/ programmes will also come about via a partnership approach (Khan 2017). Unfortunately, ODA was not taken into account while these agreements were taking place and hence, the result now is that even though CF agreements were formulated with good intentions and agreed that it would be above and additional to the ODA being provided, only about half of that additional fund has been contributed by some industrialized countries, while most are even further behind. ODA is somewhat being repackaged as CF (Oxfam 2012; Nakhooda 2013).

Attempts to Strengthen Transparency and Accountability

From COP 13 (2007) onwards, the measurability, reporting and verification of finances flowing into LDCs from Highly Developed Countries (HDCs) have been strengthened and CF's reporting guidelines were also made powerful in COPs 17 and COPs 18, under which the old industrial countries assumed obligation to report on climate finance in details both in their National Communications (NatComms) and Biennial Reports (BRs). The Common Tabular Format (CTF) for submission of CF project-related information was agreed upon as well. Later, to further strengthen and effectively create modalities and framework of transparency and accountability for mobilization of public funds, in COP 21, the SBSTA, the technical arm of the Convention, was formed. Furthermore, under the Paris Agreement (PA) of 2015, along with the establishment of Articles to enhance capacity building and create transparency framework, the need for developing nations to report their financing needs and climate funds received were also stated.

Failures in the Creation of Transparency and Accountability Modalities

Unfortunately, the status of transparency and accountability is unsatisfactory in both developed (donors) and developing (recipients) nations. Firstly, the definition creation and the resultant methodology establishment of CF is continually being hindered by the developed nations, which means there is negligible uniformity in CF accountability across countries. This, in turn, has many-a-times resulted in double/triple/quadruple counting of the funds in the books.

Hence, unsurprisingly, there is a Himalayan difference between the CF disbursed by the donors and CF received by the recipients. To elaborate, "…at COP21 in Paris, when the donors declared that they provided USD 62 bn as CF in 2014 to the developing nations, India instantly produced their research on CF showing that only USD 2.6 bn has been actually received by the DCs". Also, several NGOs, like Oxfam America have shown about 80% of CF so far delivered are ODA renamed and repackaged (Oxfam 2012; Nakhooda 2013); even worse, almost 33% of this money has been allocated to adaptation and only about 20% of it went to the most vulnerable countries, which numbers at about 100 UNFCCC Parties. Furthermore, there is a lack of granularity in project data as most countries submit compiled information, without giving project-level disaggregate data and/or explaining the financial tools used. Also, there have been reports that several of the bilateral, multilateral and international NGOs currently provide CF have creamed off a significant portion. Additionally, the accessibility of funds is also very complex, even with the establishment of a 20-member Standing Committee on Finance (SCF), which lacks in exercising its full potential towards enhancing transparency in project details and mobilization. Moreover, the projects taking place seem to be more short-term; rather than "…warranting more use of local resources and long-term investment in developing human resources and professional expertise", they are "workshop-driven".

Solving the Accountability and Transparency Issues: SBSTA Submissions

Both LDCs and HDCs have, over the years, put forth several propositions, to lower the transparency and accountability issues existent in the Climate Finance literature. To have them at one place, at its forty-fourth session, the SBSTA invited the opinions and ideas of Parties and observer organizations to be submitted to the UNFCCC Secretariat. These submissions, analysed together, will help to identify the similarities among the Parties, and those points could be a starting point for developing the modalities for better transparency and accountability for the mobilization of CF funds, instead of concentrating on unique ideas proposed by only a few nations (SBSTA 2016). Also, the similarities, if discussed, before the disparities and agreed upon to be ratified, will set a precedence for higher levels of trust among the Parties and organizational representative attending Conference of Parties—COP(s); till now, COP negotiations were hindered by donor Parties trying to proof they provided as much as their ledger said, while the recipient Parties were busy discrediting these claims, often putting the real issues—those of common methodology creation—in hindsight (Schellnhuber et al. 2010). It is to be understood that as soon the issues of accountability, transparency and lack of common methodology are solved, the Himalayan difference between CF received and CF provided will automatically become small (and may later vanish altogether) as well. The establishment of a CF definition, coupled with the recent attempts of OECD to redefine ODA (Hynes and Scott 2013) in today's context, will also be better addressed, then.

Building Bridges to Turn Differences into Strengths for Modality Creation

To better understand why a universal CF framework has still not been introduced, the differences in opinions amongst the Parties need to be analysed. Only by sorting out the differences, especially in three categories—recipient parties, donor parties and observers—can bridges be built to reach compromises and consequently better results at negotiations that benefit both the developed and the developing nations. After all, the question is of the greater good of saving the Earth's most vulnerable regions from the effects of climate change and mitigating the reasons which cause this phenomenon in the first place, not individual gains.

The Developing Nation/Recipient Parties

To summarize, from an analysis of the SBSTA submissions of the recipient parties (Table 1.1), it is evident that all of them believe that a common reporting format is currently missing, but essential to lower gaps in sent and received CF funds; this, they believe, will mitigate the accounting problem. Moreover, new instruments are needed to weigh the adaptation funds coming in, which are often confused with ODA. Several of the LDCs also point out the need for building capacity at the project proposal writing and negotiations levels among the practitioners. Furthermore, political connections should not be the bigger basis for receiving funds, but the vulnerability, governance and transparency indicators of an LDC should play bigger roles; both donor and recipient parties have roles to play here.

The Developed Nation/Donor Parties

Overall, all donors opine that the recipient parties' current proposal and report-writing acumen are flawed and need attention (Table 1.2). Also, cross-cutting projects need special attention; so does the government level activity: many encourage private sector participation for better implementation of CF projects and diversification of the proposals too. A few also highlight the problems associated with the Rio Marker instruments and the need for a better weighing mechanism.

The Observers and OECD

Brown University—This top-rated university has proposed reporting to the UNFCCC by non-Annex I Parties. They have also said that a final solution takes into account newness, "additionality" and acceptability. As reported by Stadelmann and his colleagues in 2011, "This baseline would count new sources only, meaning that only assistance from novel funding sources—such as international air transport levies, currency trading levies or auctioning of emission allowances—would be seen as new and additional." They further believe that due to the absence of baseline, billions are being spent, but without any real trust-building; this view is also supported by Khan (2014).

Climate Action Network (CAN)—They have suggested the use of electronic live reporting for real-time issue-identification and solution resulting in higher levels of transparency and accountability. Basically, they, like many others, believe in the use of online platforms and communication technology to better distribute CF and get feedback on its utilization and lacking.

World Resources Institute (WRI)—This organization has identified four issues associated with counting these finance flows including Committed versus Disbursed, Subsidy Costs (Nominal Value), Gross Flow (Net) and Total Cost (Incremental); these aggregated terms reduce complicacies, but at the same down increases accountability issues and resultantly, reduces transparency.

Table 1.1 Unique points/differences in opinions—the developing nations/recipients' perspective

Sl	Country	Thoughts/opinions
1	Brazil	Separate chapter needed on "financial resources and transfer of technology in the biennial reports"
		Accounting's definition is an "unsatisfactory" definition (which has been given in units such as AAUs, ERUs and CERs)
		Reporting should be done on more elaborate categories, project-level data provided should be more detailed
		The modalities should be developed for consideration by COP 24, with a view to making a recommendation for upcoming meetings (although this idea is not new and is already being put into action via the introduction of NatComms, BRs and the CTF)
2	Costa Rica	Proper accounting is not only necessary for mobilization, but also to meet national priorities
		For the aforementioned to happen effectively, developed countries should provide "clear and accurate information" about the funds they are providing
		The reports and communications that assess transparency like BRs and NatComms must be forward-looking (not backward-looking, like they are now)
		Transparency framework needs to be developed (Note: Bangladesh seems to be ahead of Costa Rica in this regard, as they already are implementing the previously stated climate finance transparency mechanism (CFTM) project)
3	Congo	Private finance reporting guidelines are still not there, which is hardly a surprise seeing as their roles are novel in the CF arena
		Reports focus on a lot of things and so, the accounting of financial resources are not done as efficiently and are "therefore are inadequate to support implementation of the obligation on developed country parties under Article 9(7) of the Paris Agreement" (UNFCCC 1992)
		ODA diversion must be avoided which has already been proven to be happening (Oxfam 2012; Nakhooda 2013)
		Only the grant equivalent of these instruments should be counted while reporting the financial instruments and that differentiation must occur on which type of adaptation fund is being provided
4	Ecuador	There should be summarized versions of all the submissions made by the developed country parties between 2010 and 2014 by the secretariat, especially the quantitative and qualitative elements of a pathway of CF
		Assistance for the aforementioned purpose can be taken from the syntheses of Annex I national communications
		The review process for the information provided by Each Party under Article 13.11 shall include assistance in identifying capacity-building needs, because, now, very often the really "needy" or places that actually need capacity building go into hindsight
		The identification of the type of capacity building needed must be made more efficient as, more often times than not, monetary losses occur due to a fitting problem rather than the inability to perform

(continued)

Table 1.1 (continued)

Sl	Country	Thoughts/opinions
5	Indonesia	To control the report of climate finance provided by developed country parties, involvement of donors, beneficiary country and international committee under the authority of the COP in validation the report is important; this is hardly a new idea, but the truth is, it is one that is rarely seen happening
		The development of modalities should follow the principle of common but differentiated responsibilities and respective capabilities (CBDR-RC) and take into account the country sovereignty; this means that they think, like other LDCs (e.g. Bangladesh) that the political climate and several other factors (and not only the ability to present data better), should be taken into account while providing funds
		The richer LDCs get away with more of the CF only because their negotiators, analysts and presenters are better, rather than because they are more vulnerable
6	Maldives	Short-term improvements to make the ongoing technical works of establishing definitions, creating methodologies and postulating frameworks, quicker is necessary; this is a clever plan, because often times the planners look towards the bigger picture forgetting that solving the smaller problems in the short run is more efficient as they in turn will together solve the big issue in the long run
7	Mali	Improving transparency is proposed in a vague way by mentioning climate specificity, counting both when committed and disbursed (via reporting), providing detailed information and reporting by multilateral agencies via them reporting too
8	Vanuatu	Tracking the backflow of resources to donors is needed as this increases the trust between them and the recipients and understanding the reasons will be like lessons learnt for a better future CF implementation; this is a good approach as the poorer LDCs will benefit in receiving more CF

OECD—This organization points out the importance of the Development Assistance Committee (DAC) Creditor Reporting System (CRS) which provides transparency through the collection, processing, analysis and publication of project-specific information on individual development finance activities. They opine that "modernization" of the OECD-DAC development finance framework is necessary. To that end, they are also revising the definition of ODA, so that its commonalities with CF become less pronounced and the repackaging/deviation problem reduces.

To conclude, the observer parties have evaluated and proposed the following— (1) The need for better reporting from both sides to encourage trust-building; (2) the introduction of electronic and live reporting systems (3) the introduction of a Creditor Reporting System and (4) the "redefinition" of ODA and definition of CF.

Hence, it is evident that there is a "blame game" that is being played between the donor and recipients, and hence, the negotiations often fail to bring in successful decisions. While the recipients claim that the reporting formats from the donors'

Table 1.2 Unique points/differences in opinions—the developed nations/donors' perspective

Sl	Country	Thoughts/opinions
1	Canada	Capturing the actions of other levels of government and ensuring the successful application by all providers are key to mitigating the modality challenges; this proposal, however, is a huge challenge, especially for the recipient, developing nations, because the government ministries and departments are often characterized by corruption and inefficiency
2	Japan	More privatization of CF and the relevant development of modalities are necessary
		Since SBSTA item on accounting modalities of climate finance and APA item on transparency are closely related, the details of discussions held at both bodies should be shared, and if necessary, both bodies discuss these items together; this would certainly yield better results and reduce complementarity
3	New Zealand	Donor Party reporting on the results of the CF support can be enhanced and a good way to do this is to design an effective system that improves on the existing CTFs and SCF Biennial Assessments; this would be a good step by a DC, because usually it's a "blame game" at the COPs where the donors and recipients blame each other and the LDCs fight among themselves for the limited CF; this party has decided to look at themselves before looking at the flaws of other, which is a great first step to solve transparency problems
		The modalities should reflect the mandatory and non-mandatory aspects of Article 9(7) of the Paris Agreement
		Even before deciding the contents of the framework, the countries should all agree on building it in the first place
		Developing nations should be given some flexibility in the face of their weaker capacities, which is similar to several LDC's opinions
4	Norway	Apart from the common issues of CF, other issues that need attention are the inconsistencies related to which currency exchange rates to use, which recipient countries to include in countries' reporting, whether to only include ODA or also other official flows and how to account for private climate finance mobilized
		The modalities should fit in with transformational goals, be more country specific, build upon the OECD-DAC methodology, promote harmonization among donors in terms of data and mobilization and keep better record of what the multi-laterals are doing with CF; all these smaller problems, if addressed will tone down the conflict issue between CF and ODA and also promote transparency, overall
5	Russia	Diversity in the forms of CF projects is encouraged, because it improves the financial and technical mobilization of the funds and "to utilize the UN experience to provide for systematic approach to assistance and to raise the transparency and accountability of spending the funds"
		Instead of looking towards what other donors are doing, projects should be adapted to the needs of the recipients
		Some CF should be used to finance programmes in donor states and reemphasize the importance of classification of the climate-adaptation aid projects

(continued)

Table 1.2 (continued)

Sl	Country	Thoughts/opinions
6	Slovak Republic	A common understanding about the principles of public funds mobilization is necessary, with climate finance from developed nations only to be counted, along with better monitoring, where multiple actors are involved to avoid double counting
		Designing the reporting framework to encourage and incentivize the most effective use of climate finance is encouraged; they, however, have not elaborated this point which seems to be their most important addition
7	Turkey	A concrete step would be to put a footnote for the "Sectors" column in Tables depicting the provision of public financial support: contribution through multilateral channels in 20XX-3 and the provision of public financial support: contribution through bilateral, regional and other channels in 20XX-38; this would contribute to the progress of climate finance definition issue
		The footnote should explicitly include all sub-sectors that contribute to the climate mitigation to be able to cover all parties' considerations
		The sub-sectors placed in the footnote must be the internationally accepted ones
		This idea addresses the inconsistency and granularity problems in reporting and attempts to make the reports from various donors and recipients, comparable
8	The United States	Keeping tags on cross-cutting activities is extremely important; this is really good proposal, but the problem is that these activities are quite complicated and need in-depth analysis to keep tags on
		Aggregating the finances that support capacity building and technology transfer would be better, instead of reporting on these activities in three separate tables; many negotiators are actually against the aggregation and prefer detailed tables as they provide more details, and hence improve transparency

side often contain double (sometimes triple) counting, the donors believe the gap in funds is due to the lack of report-writing capacity from the donors' sides. Additionally, poorer LDCs claim that the richer LDCs having better connections, but lower CC vulnerabilities often end up getting a larger share of CF, whereas the donors claim that the lack of ability of the poorer LDC CF practitioners and absence of project diversification, coupled with lower levels of transparency and governance and higher levels of corruption is the reason for the subsequent absence of trust-building between the donors and poorer recipients. Both Parties, however, do seem to either partially or fully agree towards the need for better reporting and weighing systems, the way forward to which is portrayed by the observer parties.

Similarities: The Starting Point for Modality Creation

Fortunately, almost all the Parties have consensus upon the need to create a common reporting format, improvement of private sector involvement and issues of public sector invention. They also encourage involvement of OECD, MDBs and IDBS, lacking currently and believe that efficient CF project formats created by other donor/recipient countries and their lessons learnt should be shared with those recipients lacking experience and knowledge in this arena. This might pose a problem though, because sharing lessons might mean the poorer LDCs will be able to claim higher shares with the vulnerability edge than their teachers (the richer LDCs); hence, donors need to take initiative to make these lessons available to all via trainings and campaigns. The similarities in opinions among several of the submissions and relevant implications are highlighted in Table 1.3.

Conclusion

Climate finance is an important part of the current and upcoming legal and policy regimes to address the changing environment and global warming. This makes improving, tracking, monitoring and reporting of climate finance extremely essential. The international treaty obligation for the developed countries to support the developing and vulnerable countries is not only about legal binding but also a matter of upholding human rights. For this, democratization of climate finance governance is imperative with core principles to be ensured through the system. These principles include accountability, transparency along with public and gender-equitable participation in the decision-making process. Furthermore, to reach a consensus, an understanding of where the gaps are occurring in opinions between the donors and the recipients is also key, to address and then effectively bridge the gaps. All of these will ensure proper mobilization of climate finance, its governance and allocation being much more systematic and transparent to have more efficient and equitable actions towards adaptation and mitigation. Also, transnational governance over climate change is crucial in the climate change regime and will only be successful in doing so when networks who act in this political sphere will collectively steer the constituents towards the public goal.

Disclaimer Readers should note that the chapter contains information updated till October of 2018 (when the chapter was submitted); later changes have not been included.

Table 1.3 Similar opinions expressed by countries via SBSTA submissions

Sl	Similar point(s)	Countries/organizations involved	Comments/implications
1	Reporting format	Brazil, Canada, Congo	Information cannot be compiled and compared as different Parties are accounting for different things without explaining what exactly they are accounting for
			Many-a-times, reports are presented in compiled manners without explaining what the components used to reach a final result meant, in the context of the country that submitted the report
			The aforementioned being practised by both HDCs and LDCs, means that the reports will not be comparable and hence, will have huge chunks of CF inflow and outflow data missing from both sides, not due to the absence of the data itself, but due to the absence of a common reporting format
		Brazil, Congo, Costa Rica, Ecuador, Indonesia, Mali, Maldives, Vanuatu, USA, Norway, Slovak Republic, Switzerland, The United States, Brown university and climate action network (CAN)	Have expressed concern over the nonexistence of a common definition of Climate Finance by UNFCCC; this lack of a common definition and resultant methodology is the primary problem in the CF literature
		Brazil, Congo, Costa Rica, Vanuatu and Switzerland	The issue of double counting is vast due to the aforementioned reasons too
2	Private sector involvement	Costa Rica, Brown university, Canada and Ecuador	The scope of CF is growing both in terms of mobilization and involvement of private parties
		Brazil, Congo, Indonesia and Maldives	There is no explanation of what is "new and addition" (Jordan and Werksmann

(continued)

Table 1.3 (continued)

Sl	Similar point(s)	Countries/organizations involved	Comments/implications
			1994) as was the status of CF under PA, 2015; this entrance of private entities into the arena, however, does show a ray of hope as private organizations are usually more adept at data collection, analysis and presentation than public ones
		Brazil, Congo, Costa Rica, Indonesia, Maldives and Russia	The expansion of the scopes of CF will not be very successful as there are issues with the reporting of mobilization of both private and public funds
		Japan, New Zealand and Norway	Expressed concerns over the "inadequate quality of data, its availability and coverage"
3	Public intervention	Indonesia, Canada and Japan	At the project level, there is high level of transparency and granularity as activities are reported with frequency and consistency in tabular formats. This was further helped on after the introduction of the CTF
		Indonesia, Canada, Japan and New Zealand	Transparency can be uplifted through public intervention. The transparency improvement project being conducted in Bangladesh by three important climate change related organizations (namely ICCAD, BCAS and C3ER) called the Climate Finance Transparency Mechanism is attempting to do just that creating an online information sharing platform so that CF funds and project data become available to the CC vulnerable public/project

(continued)

Table 1.3 (continued)

Sl	Similar point(s)	Countries/organizations involved	Comments/implications
			recipients and the platform will also invite data and information from the root levels
4	Improved involvement of OECD, MDBs and International Development Banks (IDBs)	Maldives and Vanuatu	The development of a timeline for deliverables and increased information symmetry for the recipients proposed, so that they can track the donations and adjustments in reporting parameters
		Costa Rica, Norway, Slovak Republic, Switzerland and the United States	Emphasized the role of MDBs and other multilateral donors to play a more effective role in tracking mobilization and utilizations of the CFs
		The United States and WRI	"Future modalities should allow for countries to report on how they have identified finance as being concessional or non-concessional" and determination of common approaches by IDBs and other relevant international institutions, such as the International Financial Institutions, based on the gaps already identified in their frameworks and building upon their previous experiences needs to be done, along with clarity improvement of the delivery channels and financial instruments of CF; basis of measurement must also be defined, i.e. funds committed and/or disbursed
		Slovak Republic, Switzerland and the United States	Improvements will require in-depth technical exchange among parties and transparency experts such as the OECD

(continued)

Table 1.3 (continued)

Sl	Similar point(s)	Countries/organizations involved	Comments/implications
			Research Collaborative, and OECD-DAC
		Slovak Republic, Brown University and OECD	After several consultations with, among others, MDBs and Development Finance Institutions (DFIs), the OECD-DAC revised their reporting directives on the Rio Markers to streamline guidance and take into account most recent findings from the Intergovernmental Panel on Climate Change (IPCC); the OECD Rio Markers for both adaptation and mitigation have been acknowledged as a tool for transparent and comprehensive while feasible reporting by parties in their Biennial Reports and the OECD made efforts to harmonizing and improving the applicability of Rio Markers
5	Common definitions/ formats and sharing lessons learnt	Almost all the parties, including Brazil, Indonesia, Maldives, Mali, Canada, Japan, New Zealand, Norway, Slovak Republic Switzerland and the United States	Called for establishment of a Common Tabular Format (CTF) for more consistent and transparent reporting of financial activities
		Indonesia and Japan	A way to mitigate double counting of CF would be to "invite developed country Parties should be invited to share how they have used public interventions to provide and mobilize climate finance, especially from non-public sources"
		New Zealand, Maldives and Norway	Invite developed country Parties should be invited to share how they have used public interventions to provide and mobilize

(continued)

Table 1.3 (continued)

Sl	Similar point(s)	Countries/organizations involved	Comments/implications
			climate finance, especially from non-public sources and also necessarily monitor—track and report the results
		Ecuador, Maldives, Canada and Japan, Norway and the United States	Co-operating with American Psychological Association's (2010) broad transparency framework is proposed, giving equal voice to developed and developing nations during negotiations and for accounting modalities; development, developing better mobilization methodologies and creating opportunities for workshops to teach how to track new types of CFs
		Brazil, Congo, Costa Rica, Turkey and WRI	The establishment of a common definition for CF (and appropriate methodology) will solve most of the existing challenges

References

Biennial Assessment and Overview of Climate Finance Flows (2014) United Nations Climate Change. Retrieved in September 2018 from https://unfccc.int/topics/climate-finance/resources/biennial-assessment-of-climate-finance

Bird N (2014) Fair share: climate finance to vulnerable countries. Overseas Development Institute, London

Brown J, Bird N, Schalatek L (2010) Climate finance additionality: emerging definitions and their implications. Clim Finan Policy Brief 2:1–11

Buchner BK, Oliver P, Wang X, Carswell C, Meattle C, Mazza F (2017) Global Landscape of Climate Finance 2017. Climate Policy Initiative (CPI). Derived in June 2018 from https://climatepolicyinitiative.org/wp-content/uploads/2017/10/2017-Global-Landscape-of-Climate-Finance.pdf

Chandrasekhar CP (2014) Potential and prospects for private sector contribution to post-2015 development goals: how can development cooperation strengthen engagement and results? ADB sustainable Development Working paper series No. 34

Climate Fiscal Framework (2014) Finance Division. Government of Bangladesh. Retrieved in August 2018 from https://www.climatefinancedevelopmenteffectiveness.org/sites/default/files/publication/attach/Bangladesh%20Climate%20Fiscal%20Framework%202014.pdf

CPI (2014) The global landscape of climate finance 2014. Climate Policy Initiative (CPI), USA

Hynes W, Scott S (2013) The evolution of official development assistance. Achievements, criticisms and a way forward, vol 12. OECD

Jordan A, Werksmann J (1994) Additional funds, incremental costs and the global environment. Rev Eur Int Environ Law (RECIEL) 3:81–87

Khan MR (2013) The politics of international adaptation funding: justice and division in the greenhouse. Glob Environ Polit 29–48

Khan MR (2014) Towards a binding climate change adaptation regime: a proposed framework. Routledge, London

Khan MR (2017) Climate finance: mutual accountability in use. Environment and action. Retrieved in February 2018 from www.thedailystar.net/environment-and-climate-action/mutual-accountability-use-1367128%3famp

Ludemann C, Ruppel OC (2013) International climate finance: policies, structures and challenges. In: Ruppel OC, Roschmann C, Schlichting KR (eds) climate change: international law and global governance. Nomos, Germany, pp 376–408

Morita T, Pak C (2018) Legal readiness to attract climate finance: towards a low-carbon Asia and the Pacific. CCLR, 6

Nakhooda S (2013) Mobilizing international climate finance. Lessons from the fast start finance period

Nakhooda S, Norman M (2013) Climate finance: is it making a difference? A review of options for baselines to assess climate finance pledges. Clim Dev 3:45–54

Oxfam (2012) The climate fiscal cliff: an evaluation of fast start finance and lessons for the future. Oxfam Media Advisory, Oxford

Schalatek L (2012) Democratizing climate finance governance and the public funding of climate action. Democratization 19(5):951–973

Schellnhuber HJ, Messner D, Leggewie C, Leinfelder R, Nakicenovic N, Rahmstorf S, Schlacke S, Schmid J, Schubert R (2010) Policy paper 6: climate policy post-Copenhagen. A three-level strategy for success. WBGU, Berlin

Stadelmann M, Roberts JT, Micgaelowa A (2011) New and additional to what? Assessing options for baselines to assess climate finance pledges. Clim Dev 3:45–54

Stewart R, Kingsbury B, Rudyk B (2009) Climate finance: regulatory and funding strategies for climate change and global development. NYU Press

Subsidiary Body for Scientific and Technological Advice (SBSTA) (2016) Modalities for the accounting of financial resources provided and mobilized through public interventions in accordance with Article 9, paragraph 7, of the Paris agreement. Item 13 of the provisional agenda. United Nations Framework Convention on Climate Change

UNFCCC (1992) United Nations framework convention on climate change. UNFCCC, Rio de Janeiro, Brazil

UNFCCC (2014) COP 20 outcomes. Retrieved in October 2018 from http://newsroom.unfccc.int/lima/-call-for-climate-actions-puts-world-on-track-to-paris-2015/

UNFCCC (2015) Adaptation of the Paris agreement. Retrieved on 5 October 2018 from http://unfccc.int/resource/docs/2015/cop21/eng/l09r01.pdf

Venugopal S, Patel S (2013) Why is climate finance so hard to Define? World Resources Institute (WRI). Retrieved in August 2018 from https://www.wri.org/blog/2013/04/why-climate-finance-so-hard-define

Weischer L, Wetzel M (2017) Climate project or development project: a story of definition

What is the Paris Agreement? (n.d.). United Nations Climate Change—process and meetings. Retrieved in September 2018 from https://unfccc.int/process-and-meetings/the-paris-agreement/what-is-the-paris-agreement

Chapter 2
Climate Change and State of Renewable Energy in Bangladesh: An Environmental Analysis

Kamrun Nahar and Sanwar A. Sunny

Abstract Bangladesh has also specified an unconditional contribution in its Intended Nationally Determined Contribution (INDC) to reduce harmful greenhouse gas emissions by 5% by 2030 across different economic sectors, such electric power, transportation and industry. In Bangladesh, these types of wastes are not properly utilized and result in more negative externalities. The main environmental threat from biodegradable waste is the production of methane. Biodegradable waste, when collected and processed in an industrial digester, can produce natural gas, used for homes, as well as a growing number of truck and bus fleets in developed nations. Compare this with natural gas, which contains 80–90% methane. The energy content of the gas depends mainly on its methane content. High methane content is therefore desirable. A certain carbon dioxide and water vapor content is unavoidable, but sulfur content must be minimized—particularly for use in engines. The average calorific value of biogas is about 21–23.5 MJ/m^3, so that 1 m^3 of biogas corresponds to 0.5–0.6 l diesel fuel or about 6 kWh. This overall yield of a biogas plant depends not only on the type of feedstock, but also on the plant design, fermentation temperature and retention time. In light of these joint findings, it makes upfront sense as to why direct subsidies and public financial contributions to installation costs have been crucial for the installation of some pilot plants. However, they have not provided incentives for proper and efficient operation. By contrast, the establishment of appropriate feed-in tariffs stimulates the construction of efficient plants and their continuous and efficient operation. However, besides price considerations, there remain many barriers to market penetration and development of the biogas sector.

Keywords Climate change · Renewable energy · Sustainable environment · Biogas plants · Bangladesh

K. Nahar (✉)
Environmental Science and Management, North South University, Dhaka 1229, Bangladesh
e-mail: nahar.kamrun@northsouth.edu

S. A. Sunny
Merrick School of Business, University of Baltimore, Baltimore, MD 21201, USA

© Springer Nature Switzerland AG 2021
Md. Jakariya and Md. N. Islam (eds.), *Climate Change in Bangladesh*,
Springer Climate, https://doi.org/10.1007/978-3-030-75825-7_2

Introduction

Recent research has found that there is considerably more utility-scale solar photovoltaic potential than previous estimates show, as well as lower costs ($91/MWh)
compared to coal power ($110/MWh). Additionally, excessive conversion of
cropland can likely be avoided, making it feasible also from a geospatial perspective (Shiraiski et al. 2018). In urban areas, particularly in residential and
commercial buildings, 17% of current peak demand (2 GW) can be met across the
country at $244/MWh, without any impact on croplands. At nearly 0.6 GW, wind
energy across Rangpur, Sylhet and coastal areas of Chittagong can provide around
5% of current peak demand at around over $100/MWh. Figure 2.1 shows the
spatial distribution of such renewable energy potential. Despite such potential, the
state has still planned to develop more than 13 GW of coal-powered plants 2021,
with only 2 GW of solar, even though the long-term cost-savings justify higher
investments in renewables. At this rate, and under such plans, it is unlikely that the
country will be able to meet its 2030 clean energy target of deploying 7.8 GW of
utility-scale solar. If the past is any indication of meeting future goals, the country
has fallen short of meeting the 10% renewable share by 2015, as proposed in the
signed millennium declaration. Without a centralized government agency to deal
with renewable energy issues, the feasibility of making future progress at a higher
scale is even more questionable.

Bangladesh has also specified an unconditional contribution in its Intended
Nationally Determined Contribution (INDC) to reduce harmful greenhouse gas
emissions by 5% by 2030 across different economic sectors, such electric power,
transportation and industry (Shiraiski et al. 2018). Historically, most of the progress
in renewable energy has been from bottom-up programs, such as Bangladeshi
Infrastructure Development Company Limited (IDCOL) implemented Solar Home
System (SHS) program in the early 2000s. As of last year 2017, more than
4.5 million systems, accounting for around 200 MW of electricity, have been
deployed with the help of ecosystem actors, such as NGOs and private companies.
The scheme combined innovative partnerships and financing models to deploy at
this level. The full potential might not be realized at the grid level, which only
supplies 35% of the total population, and only 3% of the population enjoying piped
gas supply. About 70% people of Bangladesh live in rural areas, where such utility
conveniences are underdeveloped—most consumers are scattered and neither grid
nor piped supply is suitable for those areas. As a result, rural-to-urban migration is
high in Bangladesh.

Drawing from such trends and perspectives, recent research on biofuels on
rooftops (Nahar and Sunny 2016) and across national croplands (Nahar et al. 2011)
allows the current research to consider an underrepresent role that Bangladesh can
play in leading low-carbon deployments (Sunny 2011, 2013, 2017). The majority of
the electricity now produced in the country is based on natural gas, which has
limited reserves and will be exhausted in the near future. To face this situation
appropriately, finding socio-technical feasible alternatives, such as renewable

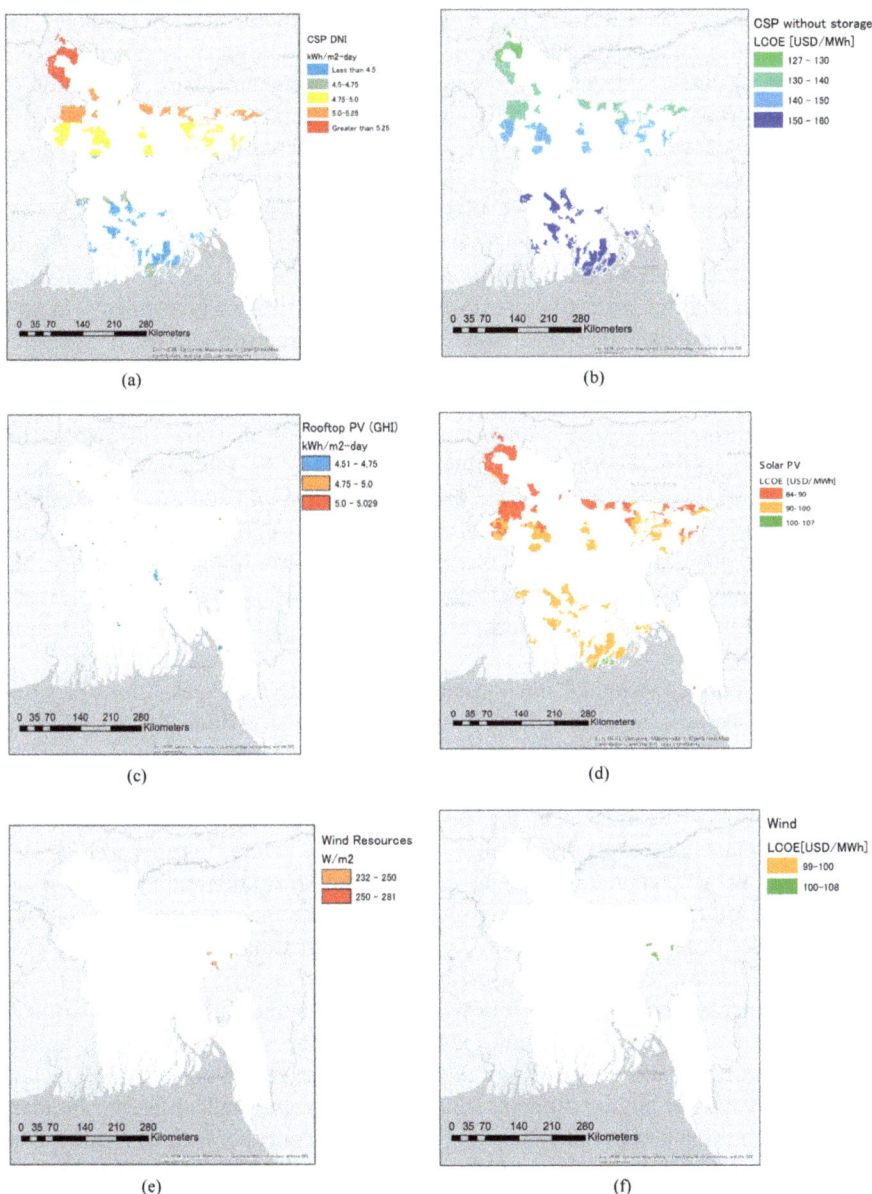

Fig. 2.1 Spatial distribution of renewable energy technology potential (adapted from Shiraiski et al. 2018) **a** Concentrated solar power (CSP) resource spatial distribution by direct normal irradiance (DNI). *Note* Low-quality or unfeasible areas are excluded. **b** Total levelized cost of electricity (LCOE) of CSP (average) without energy storage, accounting for quality, transmission line distance and road access. **c** Built environment, rooftop areas by global horizontal irradiance (GHI) for photovoltaic resources. **d** Total levelized cost of electricity (LCOE) of solar photovoltaic, (average) accounting for quality, transmission line distance and road access. **e** Spatial distribution by wind power resources. *Note* Low-quality or unfeasible areas are excluded. **f** Total levelized cost of electricity (LCOE) of wind power zones, (average) accounting for quality, transmission line distance and road access

sources of energy, is critical (Khan et al. 2014). Biofuels, such as biodiesel and biogas technology, may be one of those given that the country is suited for such production as direct replacement for petrochemical fuels and gases. The feedstock and raw materials for biogas are easily and cheaply available everywhere, as well as the local human capital aligned with maintaining and scaling such niche innovations across the regional context.

This research on biogas offers some substantial advantages over traditional methods, as there would be minimal to need of electricity supply from national grid, and even if it is needed then it will be a very small amount and for a small time. There will be minimal infrastructure development lead time or investments, as well as low-cost feedstock, given that waste and other source of fuel will be collected from the area and its periphery. With the overall cost of the proposed plant establishment being low, the generation cost of electricity from this plant can also be further lowered. Life cycle environmental costs are also low, given that the proposed biogas-based distributed power plants will be environment friendly. Building such biogas power plants for small remote area will address load shedding problem as well as provide energy to households that current lack access. Since these biogas plant will work by collecting basically cow dung and garbage, the feedstock completes the circular economy loop by using wastes as inputs and making the environment clean. See Table 2.1 for the potential levels of resource input and outputs.

With the correct technical infrastructure and incentives, excess electricity that will be produced can be supplied to the national grid or stored locally or sold to the government agencies or other nearby organizations. If able to be scaled up, the plants in large village settings may meet the demand of electricity over large area, where space used by this plant can be optimized than compared to traditional, centralized plants due to its decentralized and scalable nature. In order to judge the validity of energy potential claims, we collected data from literature on how much garbage needed to produce electricity to power a given area. From these analyses, we can direct proposals on where to set up the biogas plants and how they may meet the demand of electricity of a particular area.

The chapter is arranged in the following way. It contains a brief introduction of Bangladesh's energy potential and pathways for the next 30 years. It briefly highlights research and methodology related to biogas energy specifically, including types of biogas plants, different kinds of input waste and different boundary conditions for biogas plant efficiency. As part of this, concepts such as fermentation processes and overall mechanism of biogas production, including the role and types of bacteria in the process, their characteristics, functions, etc. are discussed. A short landscape survey was conducted, where data was collected to add nuance to the region-specific idiosyncrasies of biogas plants. From this data, we conducted some initial energy estimations at the plant level. We close with some recommendations on what can be done further to improve this proposed biogas plants.

Table 2.1 Waste (input), gas and fertilizer (output) generation (IDCOL 2013)

No of available cows	394
Dung produced per cow	20 kg/day
No of calves available	294
Dung produced per calve	5 kg/day
Average dung produced per cow head	14 kg
Total dung produced	9,350 kg
Weight of water added per kg of waste	1 kg
Slurry generated per kg of waste	2 kg
Maximum possible daily gas generation	300 CuM/day
Required dung for maximum gas generation	8,571 kg/day
Daily expected gas generation from digester	300 CuM/day
Gas output (cow dung)	0.035 CuM/kg
Required dung for expected gas generation	8,571 kg/day
Remaining dung available	779 kg/day
Dung used for digestion	8,571 kg/day
Dry matter (DM) in dung	22.5%
Slurry dry matter (SDM)	72.5% of DM in dung
Moisture content in biofertilizer after water separation (MC)	35.0%
Total biofertilizer generated	2,151 kg/day
Market price of organic fertilizer	6 BDT/kg

Institutional Collaboration for Bottom-Up Renewable Energy Projects

In Bangladesh, the first biogas plant was set up at the university campus of Bangladesh Agricultural University (BAU) in Mymensingh—a floating-dome-type plant of 3 m^3 gas production capacity. Overtime, due to the visible success of the plant, five more plants were constructed in the surrounding areas, which unfortunately did not last long due to leakage in the domes. In 1974, the Bangladesh Academy for Rural Development (BARD) constructed one biogas plant following the same design. The Institute of Fuel Research and Development (IFRD) constructed another plant in the campus of the Bangladesh Council of Scientific and Industrial Research (BCSIR) in 1976, followed by a plant at the KBM College in Dinajpur in 1980.

As the construction costs were high without the availability of subsidies, the technology did not attract mass application. It was not until 1981 that the government established the Environment Pollution Control Department (EPCD) funding a program at of BDT 3,400,000. This funded around 150 floating-dome and 110 fixed-dome plants which were developed and installed by 1984. Initial faults, such as gasholder leakage and lack of post-sale service, created a negative assessment among policy-makers and end users. During this period, various engineers received technical training support from BCSIR and constructed around 90 more plants. The Department of Livestock (DLS) also trained many of its engineers on biogas technology and constructed about 70 plants by 1990.

Social movement actors, such as local NGOs like Danida, BRAC and Grameen Bank, took further initiatives to popularize the technology, with support from BCSIR. Almost 250 plates were constructed across Bangladesh. Under the "Fuel Saving Project" implemented from 1989 to 1991, IFRD trained local youths who constructed almost 130 floating-dome type biogas plants on the premises of affluent farmers, supplying the gasholder free of cost. Again, these plants also did not last long. LGED constructed a floating-dome model biogas plant in Kurigram district and arranged a seminar there on December 27, 1986—joined by around 300 scientists, experts, engineers, politicians and persons interested in biogas technology. Until 1992, they constructed more plants following the same design, and although they initially worked successfully, after five or so years, the plants failed due to leakage in the gasholder, pipeline and burner.

In 1989, one scientist from BCSIR received training on biogas technology from Biogas Research and Training Centre (BRTC), in China. After returning, he constructed one fixed-dome Chinese model biogas plant at BCSIR campus which exists to this day. Following the success of this design, another LGED engineer replicated the technology and constructed two biogas plants in Noakhali in 1992, with financial support from Danida. These plants are also still in operation (as of 2017). Given this success, LGED engineers and BCSIR scientists collaboratively constructed about 50 fixed-dome-type biogas plants in different districts at the cost of the users, and thus the fixed-dome model biogas plant was successfully introduced in Bangladesh.

Following this success, under Slum Improvement Project and Secondary Towns Infrastructure Development Project of LGED collaboratively constructed about 100 plants during the period 1992–1996. A MoU was signed between BCSIR and LGED in 1993, training more engineers. By 1994, LGED supported the establishment of an ecological village in Amgram, in Madaripur district constructing 15 domestic biogas plants using night soil, kitchen waste, water hyacinth, etc. A year later, the first phase of Biogas Pilot Plant Project implemented by BCSIR, introduced a further 4,664 fixed-dome plants were constructed throughout the country before its end in 2000. BCSIR began employing applied scientists and trained 128 diploma civil engineers who were assigned responsibilities for motivation, installation, and after sales service throughout the country—the initial missing piece that contributed to failure and lack of uptake across the masses. In addition, 898 youths were trained to support the project as a labor force. The biogas farmers received an investment subsidy of BDT 5,000 under the project to further incentivize demand-side adoption.

Besides leading the technical development, BCSIR enacted agreements such as MoUs to partner with other institutions, such as BRAC, LGED and DLS for research, training and dissemination of the biogas technology. Initial outcome assessments reported that, by 1999, 99% of the plants installed under the project were in operation, while 91% of the owners could meet their household fuel demand through biogas. Given the economic success of the deployments, stake-holders further adjusted the business model by extracting the bioslurry from the biogas plants for use in horticulture, pisciculture and agriculture. The average saving per plant amounted to BDT 759 per month. The second phase of the project began in mid-2000 and continued for four years.

The target for this phase was set at 20,000 biogas plants, out of which 17,194 plants were finally built. The investment subsidy for the owner was also increased to Taka 7,500 per plant. In addition to the diploma civil engineers employed and paid on a monthly basis by BCSIR, an agency system was introduced on incentive basis—an example of another co-creative business model innovation. This phase enacted a wider deployment scheme, identifying around 50 agencies in different areas who would receive a fee of BDT 5,000 per plant as service charge. About 128 engineers, 100 masons and 250 local youths were trained for 10 days under the project.

With the success of this project, a further 5,000 domestic goal was set by the Minister of Agriculture, in collaboration with LGED. By mid-2003, around a 1,200 biogas plants were completed, with the subsidy increasing from BDT 5,000 to 7,500. In 1994, the government of Bangladesh created Infrastructure Development Company (IDCOL) with financial assistance from the World Bank to support all kinds of infrastructure development, with focus on energy related infrastructure, finally able to successfully establish 450 MW power plants at Meghna Ghat, installing the 125,000 solar home systems. With support from Netherlands Development Cooperation (SNV), a project for the extension of biogas technology in Bangladesh is launched. Under this program, more than 36,000 biogas plants were planned by 2009 implemented by a 16-organization partnership. The move-ment was further legitimized by Grameen Shakti (GS), a member of the Grameen family, incorporated in 1996 as a "not for profit company" to promote, develop and popularize renewable energy technologies in the remote rural areas of Bangladesh. With a rapid scale capability, GS staffed around 800 engineers with 1,000 field staff, serving more than 600,000 beneficiaries through its 300 unit offices, under 44 regional offices under and 6 divisional offices.

GS brought legitimacy to the movement, overall installing more than one hun-dred thousand solar home systems, for which they received several international awards from USAID, the Prime Minister of Bangladesh and the 2006 Ashden Award from UK. By early 2005, they launched a program for the extension of biogas technology in the country. Until June 2007, they had constructed more than 1,000 biogas plants in different districts of the country. Grameen Shakti has two systems for the biogas extension program—for small family-size biogas plants, i.e., 1.6–4.8 m^3 gas production per day, they give subsidy of Tk. 7,000; and for bigger size plants, they do not give any subsidy.

To spearhead innovative financing models, for any biogas plant, GS provides loan without collateral and recovers it in 24 equal monthly installments with 6% service charge. Recently, Grameen Shakti has been providing technical and financial support for the generation of power using biogas. Instead of developing a new type of generator, GS is using the available gas, petrol and diesel generators for producing electricity. Although biogas technology is getting increasing attention all over the world, the use of bioslurry is still neglected. Giving priority to the bio-slurry, IDCOL renamed their project title, and Grameen Shakti appointed one expert exclusively for the promotion of bioslurry. But still, in most cases, bioslurry is not properly used. It is mainly because of the ignorance of the farmers. Recently, Grameen Shakti has taken a decision to demonstrate use of slurry in different districts, so that the farmers become interested in using bioslurry. In order to explore the technological potential of these types of bottom-up renewable energy projects with institutional support, it is important to briefly outline the energy system and then conduct an economic analysis. The brief outline below will be conducted from both a chemical and mechanical perspective, before evaluating the economic thresholds for future institutional support discussions.

Energy Technologies—From Inputs to Outcomes

In order to judge the efficacy of an efficient renewable energy generation system that performs across contexts in economic ways, it is vital to understand the technology—its inputs, processes, outputs and outcomes. In Bangladesh, there are different types of waste and proper using of these types of waste electricity will be produced and will be used some areas. In this chapter, the theory part will discuss different types of biogas power plant, different kinds of biogas waste like biodegradable, non-biodegradable, organic and inorganic waste. Many examples are given also for different types of waste. Organic matters like cow dung, poultry litter, human waste, city waste, water hyacinth, agriculture waste, etc., if decom-posed in an anaerobic condition at certain range of temperature, produce biogas. Anaerobic digestion is a phenomenon by which organic matter is transformed into methane (CH_4) in the absence of air. It is combustible like natural gas, odorless, smokeless and invisible in bright daylight. It composes of mainly methane (CH_4) and carbon dioxide (CO_2) with a little trace of some other gases like hydrogen (H_2), nitrogen (N_2), oxygen (O_2), etc. (Gofran 2012). When human excreta and poultry litter are used as raw materials, little existence of hydrogen sulfide (H_2S) is observed. Biogas sources can be classified in many ways, for example, by origin, or by different characteristics and properties (Launder 2002). Biogas sources can be divided into biodegradable wastes, sewage wastes, organic municipal solid waste (MSW) and other plant-life and aquatic energy crops, such as water hyacinths (Nahar 2012).

Biodegradable waste or biowaste is defined as biodegradable garden and park waste, food and kitchen waste from households, restaurants, caterers and retail premises, and comparable waste from food processing plants. It does not introduce forestry or agricultural residues, manure, sewage sludge or other biodegradable waste such as natural textiles, paper or processed wood. It also excludes those by-products of food production that never become waste. The list of biodegradable wastes includes leaves, wet kitchen waste, animal or human excrement, vegetable and food wastes, papers and certain plastics. These types of waste can be found in municipal solid waste (sometimes called biodegradable municipal waste, or BMW) as green waste, food waste, paper waste and biodegradable plastics.

In the absence of oxygen, much of this waste will decay to methane by anaerobic digestion. Using biodegradable waste, it is possible to generate electricity. In Bangladesh, these types of wastes are not properly utilized and result in more negative externalities. The main environmental threat from biodegradable waste is the production of methane. Biodegradable waste, when collected and processed in an industrial digester, can produce natural gas, used for homes, as well as a growing number of truck and bus fleets in developed nations. An even more useful by-product, especially for an agrarian community, is that after producing a low-emission renewable fuel, the remaining solids and liquids are a perfect fertilizer and compost, which farmers often purchase at high costs. Sewage waste is a source of biogas that is comparable to the other animal wastes. Energy can be extracted from sewage using anaerobic digestion, drying and incineration. Organic MSW is any matter collected from commercial or residential properties such as food waste and paper. Organic waste, whether from commercial or residential properties, makes up a substantial amount of waste that is land-filled. As with other wastes, it can be converted into energy by various ways. One is direct combustion (incinerator), or by anaerobic digestion in a land-filled or in a process plant.

Various agriculture outputs, beyond residential and commercial wastes, also present opportunities as feedstock for biomass-based energy generation. At a slaughterhouse or a fish-processing plant, there is often a large amount of organic waste. This has the possibility of being a danger to the environment and human or animal health, with costly complaint movement. Various policies specify these animal wastes may be disposed of safely with an opportunity to be used as a feedstock for anaerobic digestion. For example, several options for collecting and storing cattle manure are available, depending on the manure form. Common storage methods include under floor pits, outdoor (above or below ground) structures, earthen pits, lagoons and holding ponds. Flushing gutters and scraper systems are among the methods used to collect and transport manure to appropriate storage facilities. Cattle farming techniques significantly affect the quantity and quality of manure that may be delivered to the anaerobic digestion system. The number of cows, the housing, transport and bedding systems used by the farms determines the amount of slurry that must be used and therefore the amount of energy produced. Cattle farming may be housed using a variety of methods. The most commonly used systems include free stalls, corrals with paved feed lanes and open lot systems.

Overall, animal wastes, such as manures, renderings and other wastes from livestock finishing operations, although contain energy, are primarily motived not for biogas processing of animal wastes for energy generation, but for the mitigation of disposal. This is especially true for animal manures which are typically disposed of through land application to farmlands. Tightening regulations on nutrient management, surface and groundwater contamination and odor control are beginning to force new manure management and disposal practices. Biogas technologies present attractive options for mitigating many of the environmental challenges of manure wastes. The most common biogas technologies for animal manures are combustion, anaerobic digestion and composting. Moisture content of the manure and the number of contaminants, such as bedding, determine which technology is most appropriate. Certain commercial industries, such as dairy, are particularly well-suited to biogas-to-energy opportunities because of the large volume of manure that its operations produces.

Environmental Processes

Biogas typically refers to a gas produced by the biological breakdown of organic matter in the absence of oxygen. Organic waste such as dead plant and animal stuff, animal dung and kitchen waste can be converted into a gaseous fuel called biogas. Biogas originates from biogenic material and is a type of biofuel. It is produced through fermentation of organic matter in an anaerobic condition—a process that not only produce gas, but also kills all harmful bacteria. All the hazardous materials that pollute the environment or risk the local community health through various diseases could be potential input raw materials for a biogas plant. At the same time, it gives valuable organic fertilizer at the point of consumption for most rural areas that have an agrarian societal landscape. This decentralization is especially appropriate for the rural areas, where neither piped gas supply nor grid electricity supply is prevalent. Given that biogas can meet the need for both gas and electricity, such forms of energy have multiple advantages. In a country like Bangladesh, where waste to energy has immense potential to generate electricity, biomass–biogas holds much untapped promise and technological potential that can be beneficial in many ways—from eco-friendly waste disposal, rural income, and also for energy generation.

The liquid, which carries different species of biogas microbes that performing the function of degrading anaerobically organic substance to yield methane, is called inoculums. Natural biogas fermentation is impossible in absence of enough number of biogas microbes. As for example, 1 ml biogas fermentation liquid contains 10^6–10^8 methanogenic bacteria those are sensitive to environmental condition. Usually, the number of seeding bacteria in a fresh material fluid is bellow this standard; therefore, enough inoculums must be added to it during the startup of fermentation (Zehnder and Mitchell 1978). In the fermentation process, different bacteria play different role according to their nutrient requirement. Bacteria for biogas convert complex organic compounds into methane generally in three steps. In every step, some additional bacteria provide the impetus for this conversion. Their functions in different steps of biogas production include activity of hydrolytic, acetogenic and other methane-producing bacteria. Only a few numbers of substrate used by methanogenic bacteria to produce methane are researched—hydrogen, carbon dioxide, acetic acid, formic acid, methyl alcohol and methylamine. In fact, among these components, hydrogen, carbon dioxide and acetic acid are used to produce methane.

In practical inoculum of different sources contain different colonies of biogas microbes and each of them acts upon some particular fermentation material most efficiently. The organic sediment from the hole of compost, ponds and sewerage, the drain of the slaughterhouse, sewerage of butchery and sewerage of food processing factories are the examples of ideal inoculum. Liquid ingredients of waste of domestic animals are used as an inoculum. In the case of fermentation liquid of a biogas digester uses as an inoculum of another digester, amount of inoculum may be 30% of total fermentation liquid and in the case of suing as sediment inoculum, the amount of inoculum may be 10% of total liquid. The sludge taken away from the sewage treatment plant near the city can be used as inoculum for biogas fermentation in big- or medium-size biogas digesters. The sieve thick liquid of cattle dung and horse dropping can be used as inoculum as well. The amount of inoculum may be as high as the fermentation process may start quickly and time for organic compound to converted into methane so quickly. Fermentation temperature of around 35 °C, with the length of fermentation period being 60 days for the excrement material and 90 days for stalk type. As the ideal temperature for biogas is around 35°, the temperature in Bangladesh usually varies from 6° to 40°. But the inside temperature of a biogas digester remains at 22°–30°, which is very near to the optimum requirement.

Biogas additives is one kind of substance, a small amount of which applied to the fermentative liquid, and then biogas production product increases at a mentionable rate. There are different types of biogas promoter such as enzymes, inorganic salts, and especial organic and inorganic substances. For instance, if some cellulose is mixed with fermentation liquid, degradation process of cellulose substances is sped up. When 5 ppm of rare earth element (R_2O_5) is added to fermentation liquid, then biogas production increases by about 17%. In the case of biogas production from crops stalk-type substances too, definite amount of nitrogen-type fertilizer such as ammonium hydrogen carbonate is giving to it and for this biogas production increases at a mentionable rate. Although not environmentally preferred, the addition of active carbon or coal increases biogas production. Moreover, by flowing hydrogen gas into fermentation liquid, methane production is increased. At the same time, the substances which are present in the fermentation liquid and which can inhibit the biogas production are called biogas inhibitor (resistant). Different types of organic and inorganic substances such as metal ions, salts, different bacteria and synthetic chemical substances are included in this category. As for example, if the amount of ammonia nitrogen (probably ammonium nitrate) is 1,500 ppm or that of acetic acid is 2,000 ppm in fermentation liquid, the production of biogas can be stopped. But, according to many scientists, biogas bacteria are enabling to deactivate these all biogas resistant.

Hydraulic retention time (HRT), i.e., the period for which the raw materials remain in the digester, is fixed on the basis of economic rate of gas production period of the raw materials. The retention time is calculated by dividing the total volume of digester by the volume of inputs added daily. The retention time is also a function of the type of input and the ambient temperature. In IDCOL model, the hydraulic retention time (HRT) is fixed at 45 days. Experiments and experiences show that cow dung, human excreta and poultry droppings produce gas at economic rate in the climatic condition of Bangladesh up to a period of 45 days. The loading rate is also important, given that the size of the digester is fixed on the basis of availability and nature of raw materials. In case of cattle dung, sometimes, it is seen that the cattle remain outside for the whole day and the dung is not available for use in the digester. In such case, only the dung available at nighttime should be taken into consideration. After construction of the plant, size of the digester is fixed, and the loading rate should be maintained as per requirement of the digester. Both overloading and underloading affect the gas production.

The water ration is also to be noted; as generally, cow dung contains 17% solid and poultry droppings contain 30% solid. Experiments show that, for better gas production, the solid content in the mixed slurry should be 8%. It is suggested that, to maintain this ratio, in case of cattle dung water should be mixed equal to the volume of the raw materials and in case of poultry droppings, water should be double. Relatedly, carbon–nitrogen ratio acts as an important factor in producing biogas. Materials with different carbon–nitrogen ratio differ widely in their yields of biogas. Practice shows that the result of fermentation will be good, if the carbon–nitrogen ratio of materials ranges from 20:1 to 30:1. In human waste, percentage of carbon is 2.5 and that of nitrogen is 0.85, i.e., the carbon–nitrogen ratio is 6:1. Domestic waste consists of different types of materials, and their carbon and nitrogen content also differ. In nobody slum, carbon and nitrogen in the domestic waste was found to be 14 and 0.54, respectively, i.e., the carbon–nitrogen ratio is 27:1, which is suitable for biogas production. So, by mixing domestic waste with night soil, efficiency of the mixture is improved.

To indicate the biogas-producing rate of the material, the total solid (TS) contained in a certain amount of materials in a biogas digester is usually used as the material unit. The total solid of a material includes two parts, volatile solid (VS) and ash content. The latter is the inorganic part of the total solid that is difficult to decompose and cannot be converted into methane in the process of fermentation. Biogas fermentation requires a certain range of material (TS) concentration. This range is rather wide, usually from about 1 to 30%. Even if it is higher than 30%, the production of biogas can still occur. Experiments show that, at different fermentation temperature, the percentage of total solid in a biogas mixture should be around 8. It is difficult to maintain this proportion in night soil. Because, a man usually leaves about 0.3 kg of night soil but use about 20 kg of water for clinching. Thus, the proportion of solid comes down to about 0.26%. In such case, no gas can be expected. In slum area, specifically in slum areas, due to shortage of water, people generally use 3 kg of water for clinching, which gives TS 1.6%. In such case, optimum production cannot be achieved, but 60–70% efficiency can be expected. In nobody housing, BSCE is using domestic waste with night soil. The domestic waste contains about 25% TS. As a result, the percentage of TS in the mixture increased to about 6%, which is very near to optimum requirement.

While biogas technology, the generation of a combustible gas from anaerobic biomass digestion is a well-known technology with millions of biogas plants in operation throughout the world with uses ranging from the incoming gas for direct combustion in household stoves or gas lamps, and producing electricity from biogas is still relatively rare in most developing countries. Biogas can be used in similar ways to natural gas in gas stoves, lamps or as fuel for engines. It consists of 50–75% methane, 25–45% carbon dioxide, 2–8% water vapor and traces of O_2 N_2, NH_3 H_2 H_2S. Compare this with natural gas, which contains 80–90% methane. The energy content of the gas depends mainly on its methane content. High methane content is therefore desirable. A certain carbon dioxide and water vapor contents are unavoidable, but sulfur content must be minimized—particularly for use in engines. The average calorific value of biogas is about 21–23.5 MJ/m^3, so that 1 m^3 of biogas corresponds to 0.5–0.6 l diesel fuel or about 6 kWh. This overall yield of a biogas plant depends not only on the type of feedstock, but also on the plant design, fermentation temperature and retention time. Certain feedstock, such as maize silage, yields about 8 times more biogas per ton than cow manure, but also costs more. About two livestock units (corresponding to about 2 cows or 12 rearing pigs) plus 1 ha of maize and grass are expected to yield a constant output of about 2 kWel (48 kWhel per day). In the context of South Asia such as Bangladesh, a typical specific input–output relation of about 10 kg of fresh cattle dung (the approximate production of one cow on one day) plus 0.06 l diesel fuel to produce 1kWh electricity. Table 2.2 summarizes some values for fuel and gas replacement through biogas-based generation.

Theoretically, biogas can be converted directly into electricity using a fuel cell. However, very clean gas and an expensive fuel cell is necessary for this process. As a result, more R&D is required for this technology to scale in developing nation contexts. For the most part, biogas is used as fuel for combustion engines, which convert it to mechanical energy, powering an electric generator to produce electricity. Appropriate electric generators are available with the underlying technology being well known with simplicity in maintenance. Even universally available three-phase electric motors can be converted into generators.

As a result, in theory, biogas can be used as fuel in nearly all types of combustion engines, such as gas engines (Otto motor), diesel engines, gas turbines, Stirling motors, etc. However, technologically this stage remains far more challenging—that is, the combustion engine using the biogas as fuel. Often times, gas turbines are used as biogas engines, due to their small, modular and strict exhaust emissions requirements compliance. Small biogas turbines with power outputs of 30–75 kW are commercially available but are rarely used for small-scale applications in developing due to cost. Furthermore, due to their spinning at very high speeds and the high operating temperatures, the design and manufacturing of gas turbines is challenging, and maintenance requires specific skills which puts its adoption at odds with the local labor force able to service them. External combustion engines such as Stirling motors have the advantage of being tolerant of fuel composition and quality. At the same time, they are limited to a number of very specific applications due to being both expensive and having low efficiency. In most commercially run biogas power plants today, internal combustion motors have become the standard technology either as gas or diesel motors.

Table 2.2 Power generation and fuel replacement capacity (IDCOL 2013)

Existing generator	
Size	62.5 kVa
Diesel requirement of generator (input)	6 L/h
Diesel required (electricity output)	0.25 L/kWh
Proposed generator	100 kVa
Power factor	80%
Efficiency	80%
Generator rated power	64 kW
Gas required per unit of electricity	0.53 CuM/kWh
Maximum running hours per day	10
Total daily electricity generation	640 kWh
Total daily biogas requirement for electricity	339 CuM
Net calorific value of diesel	43,400 kJ/kg (36,923 kJ/L)
Net calorific value of natural gas	950 BTU/cft
Methane content in natural gas	95%
Methane content in biogas	60%

End-Use Consumptions

In most commercially run biogas power plants today, the internal combustion motors have become the standard technology either as gas or diesel motors. They have quite high requirements regarding the fuel quality. The gas is used as the fuel for a combustion engine, converting it to mechanical energy which powers an electric generator to produce electricity. Electricity production from biogas can be a very efficient method for producing electricity from a renewable energy source. However, this applies only if the emerging heat from the power generator can be used in an economically and ecologically sound way. The average calorific value of biogas is about 21–23.5 MJ/m^3, meaning that 1 m^3 of biogas corresponds to 0.5–0.6 l diesel fuel or an energy content of about 6 kWh. However, due to conversion losses, 1 m^3 of biogas can be converted only to around 1.7 kWh. Bigger biogas plants are generally more cost-efficient than smaller ones. However, electricity generation from biogas is a technology appropriate even for relatively small applications in the range of 10–100 kW. For smaller scale use, the design of an electric generator is similar to the design of an electric motor. Most generators produce alternating AC electricity; they are therefore also called alternators or dynamos. Electric generators are virtually available in any country and in all sizes. The technology is well known, and maintenance is relatively straightforward. In most cases, even the universally available three-phase electric motors can be converted into generators, and rural mechanics have a less steeper learning curve for maintaining such assets.

Most engines originally intended for cars, trucks, ships or stationary use can run on biogas as fuel and are available almost everywhere within a power range between

10 and 500 kW. This holds true especially in the case of dual-fuel use. Robust engines with a certain sulfur resistance are mostly free of non-ferrous metal with a copper content as these materials are very susceptible to the damage through sulfur rich biogas. In theory, biogas can be used as fuel in nearly all types of combustion engines (Cheong 2005). For example, it can be used in a Stirling motor, wherein biogas is combusted externally, which in turn heats the Stirling motor through a heat exchanger. The gas in the Stirling motor expands and thereby moves the mechanism of the engine, with this resulting work being used to generate electricity. Such engines are, however, quite expensive and are characterized by low efficiencies which is why their use is limited to a number of very specific applications.

Within internal combustion engines, diesel engines are also candidates that operate on biogas, but only in dual fuel mode. To facilitate the ignition of the biogas, a small amount of ignition gas is often injected together with the biogas. Almost every diesel engine can be converted into a pilot injection gas engine as these motors can run in dual-fuel mode and have the advantage of using gas with low heating value. However, they also consume a considerable amount of diesel. At up to 200 kW engine sizes, the pilot injection engines seem to have advantages against gas motors due to slightly higher (3–4%) efficiency and lower investment costs. On the other hand, gas motors with spark ignition (Otto system) can operate on biogas alone, although in practical use, a small amount of petrol (gasoline) is often used to start the engine. This technology is used for very small generator sets, of around 0.5–10 kW, as well as for large power plants.

For use in gas or diesel engines, the gas must fulfill certain requirements, such as the methane content being as high as possible given it is the main combustible part of the gas. Additionally, the water vapor and CO_2 content should also be as low as possible, mainly because they lead to a low calorific value of the gas. This water vapor content can be reduced by condensation in the gas storage or on the way to the engine. Finally, the sulfur content, mainly in form of hydrogen sulfide (H_2S), must be low, as it is converted to corrosion-causing acids by condensation and combustion. The reduction of the hydrogen sulfide (H_2S) content in the biogas can be addressed via a range of technical (chemical, biological, or physical) methods, both internally and externally. Additionally, given that complete elimination is generally unnecessary for use in robust engines, an optimized steady fermentation process with continuous availability of appropriate feedstock is important to produce a gas of homogenous quality. At times, the injection of a small amount of oxygen (air) into the headspace of the storage fermented leads to oxidation of H_2S by microorganisms, thereby eliminating a considerable part of the sulfur from the gaseous phase, which is the most frequently used desulfurization method (95% Sulfur content) due to its cost-effectiveness. Another way to externally treat in a filer is by enacting as the active agent of iron hydroxide: $Fe(OH)_2 + H_2S$ $\rightarrow FeS + 2H_2O$ for a process reversible, regenerating the filter by adding oxygen.

Gas turbines are also occasionally used as biogas engines given their modular size and meeting strict exhaust emissions requirements, mostly for operation on landfill and digester gases. Small biogas turbines with power outputs of 30–75 kW are available in the market, although their use is rare for small-scale applications in developing countries. They are expensive and due to their spinning at very high speeds and the high operating temperatures, the design and manufacturing of gas turbines is a challenging issue from both the engineering and material point of view. Almost 70% of source potential is from agricultural sources, with 7% each from commercial or wastewater sources, 6% from residential and 12% from landfill gases. Maintenance of such a turbine is very different from well-known maintenance of a truck engine and therefore requires specific skills, which are often outside of the capability of local mechanics in a developed setting.

Economic Aspects

Economically, electricity from biogas must compete with electricity generation from fossil fuels, but also from other renewable energies. Some of the supporting factors are the rising prices of fossil fuels, compounded with the low reliability of electricity provision from national grids with persistent risk of power cuts and vulnerability of other renewable energy types due to feedstock issues. At the same time, relatively low prices of fossil fuels and the unfavorable conditions for selling electricity at the retail rate or given infrastructure make up some of the inhibiting factors. Furthermore, being very investment-prone, actors often need to buy high quality components from industrialized countries and advanced markets, compounded with the lack of general awareness, and human capital issues, such as capacity and experience preventing the economic operation of such high-end infrastructure components (Wikberg et al. 1998).

The economic feasibility of a biogas plant depends on the economic value of the entire range of plant outputs. Some general considerations include the electricity or mechanical power of biogas feedstock yield and conversion, the related heat, from co-generation within the combustion engine. Some detailed inputs, such as the sanitation effect with COD and BOD (chemical and biological oxygen demand) reduction in the runoff of agro-industrial settings, and the feasibility of using the slurry used as fertilizer. In order to match the optimal conditions, most of the commercially run biogas power plants in developing contexts are of medium size and are installed in industrial settings. Within these, they primarily use organic waste material from agro-industrial production processes such as animal manure, slaughterhouse waste, or residues from food processing. Assessments of economic feasibility are contradictory or inconsistent. Many press releases and information from biogas power plant producers refer to payback periods of only 1.5–2.5 years. In such cases, the electricity from biogas plants can be compared to the price of electricity provided through the national grid or the price of bottled LPG. However, these figures are unrealistic, except for direct thermal energy use as for cooking energy, or in very few locations with extremely expensive diesel fuel.

We can conduct energy estimation from biogas feedstock, with an overall appraisal of energy production potential, technical and economic aspects of energy —such as conversion to electricity and end-use consumption, such as the appropriate combustion engines and gas quality. We also discuss the potentials, obstacles and necessary framework conditions for the utilization of biogas for small- and medium-scale electricity generation in developing contexts. From an economic standpoint, the technology has been available on the global market, during which time the technological difficulties were confronted and resolved. For example, different methods of desulfurization have been successfully established and combustion motors tolerant to biogas that have proven their durability are available in the market. Sufficient know-how for planning and constructing reliable biogas power plants is also available, although the construction of efficient and reliable biogas power plants, at least some technical core components must be imported from industrialized countries (Ghimire 2005). The electricity generation component of a biogas power plant does not require much more know-how and effort for maintenance than a normal generator set for fossil fuels with a well-functioning biogas fermentation process as an indispensable prerequisite.

For completion the design of an optimum model of a biogas-based efficient energy generation system, some initial surveys were conducted in different locations to correspond to secondary data collected from additional records across locations references. It was revealed from the user survey that the users of 1.2 and 1.6 m^3 plants feed all their available raw material into the plants, which meet the required quantity. On the other hand, the users of 2.0–4.8 m^3 size plants avail higher quantity of raw material than the required quantity (IDCOL 2013). The users of 2.0 and 2.4 m^3 size plants utilize a little higher quantity than required/recommended. No variation and/or minor variation in the required/recommended quantity and the quantity of raw material actually used per day indicate good operation skills of the plant users. But the use of less quantity compared to what is required indicates users' (3.2 m^3 plants) lack of knowledge.

Table 2.3 shows that the 99% of the households, the main reason for construction of a biogas plant was to produce gas for cooking. Other major reasons include reduction of cooking time and hygienic reason that would reduce smell and improve the environment of household. It is also important to note that about 72% of the households expected reduction of energy cost, whereas 75% was interested in production of improved bio fertilizer from the slurry. Additional income from sale of biogas to neighbors, or production of electricity from biogas, was not of consideration for most in their decision to install a biogas plant.

In light of these joint findings, it makes upfront sense as to why direct subsidies and public financial contributions to installation costs have been crucial for the installation of some pilot plants. However, they have not provided incentives for proper and efficient operation. By contrast, the establishment of appropriate feed-in tariffs stimulates the construction of efficient plants and their continuous and efficient operation. However, besides price considerations, there remain many barriers to market penetration and development of the biogas sector. Most importantly, the overall lack of awareness of biogas opportunities still plague mainstream economic and scientific actors in the nation. The high upfront costs for potential assessments and feasibility studies and perceived lack of access to appropriate finance, in addition to underdeveloped local capacity for project design, construction,

Table 2.3 Primary household motivations for constructing biogas plants

Types of reason	No. of respondent	%
Produce gas for cooking	207	99.00
Production of electricity	10	4.78
Reduce the cooking time	185	88.5
Production of improved biofertilized	157	75.1
Reduction of smell from environment	168	80.4
Reduce energy cost	151	72.3
To improve the environment of house	167	79.9
To earn additional money	16	7.66

Note Percentage does not add up to 100 due to multiple responses

operation and maintenance provide problematic barriers. For a developing nation, the legal framework is still immature and certain institutional conditions can still complicate alternative energy production and commercialization. For example, without a net-metering or robust feed-in-tariff, the right to sell electricity at local level is non-existent for small players. As long as the national framework conditions are not favorable, electricity generation from biogas will remain limited to a few pilot applications.

Conclusion

Electrical energy plays a vital role in development of a country. The advancement of a country is measured in terms of per capital consumption of electrical energy. Although biofuels may not solve all of the current power crisis, by incorporating it into the mix, it is possible to minimize the crisis by using the proposed renewable electrical systems. With increasing industrialization and urbanization in Bangladesh, the overall demand for natural gas, which power most of the economy, will continue to grow. It is said that the country would require about 13.6 tcf of gas up to 2020, about 26.7 tcf up to 2030 and about 62 tcf up to 2050. With natural gas as the single significant commercial energy resource available in the country, it appears that the present reserve of 11.6 tcf may not run beyond 2020, with solar and wind at a significant institutional disadvantage. Hence, a gaseous substitute should be feasibly sourced, which can be decentralized and managed by the local work-force. Low-income developing countries like Bangladesh are very much susceptible to the setbacks arising from the ongoing energy crisis. Natural gas lies at the heart of the country's energy usage. Such an overwhelming historic dependence on biomass has brought into focus the substantial amount of renewable energy resources available in the country. While scholars tout the technology potential of localized and non-exhaustive sources of energy in the form of solar, biomass, nuclear, hydro and wind, socio-technical systemic research and regional historicity gives a slight advantage to harnessing biomass-based power, such as biofuel and biogas, in order to provide an environmentally sustainable energy security (Mondal et al. 2010). Biogas is the most common form of renewable energy in Bangladesh, where the populace, give their agrarian genealogy, have some knowledge.

Often the rural population had to historically rely on the traditional biomass sources for household supply of energy, which for an agricultural country like Bangladesh makes sense given that biomass, such as cattle dung, agricultural residue, poultry dropping, water hyacinth, rice husk, etc. is available in huge amount. With the decade-long institutional work done by collaborative public, non-profit and private actors who have introduced considerable amounts biogas plants in rural areas across the country through initiatives with partner organizations such as National Domestic Biogas and Manure Program (NDBMP), more investments, both financial and institutional need to take place to carve a landscape for biomass. Most importantly, in rural regions, where the marginal utility of electricity is higher for welfare, can now be provided with affordable power supply. Moreover, the use of renewable energy sources will become increasingly necessary, if collectivity, the world is to achieve the changes required to address the impacts of global warming. Biogas has potential as various nations are setting goals since the technology is proven. The world's progress toward a healthy increasing in renewable energy use is evident in such individual goals set by numerous countries and subsequent rapid growth and development. Recently, (2014) Belgium has established latest biogas power plant on the world which can produce 215 MW of electricity and heat sustainable biogas, being 100% of raw materials biomass, such as wood chips and agro residues. Bangladesh itself has high potential of energy crop feedstock production (Nahar and Sunny 2011; Nahar 2011) such as Jatropha, sweet sorghum, castor, etc. (Nahar and Ozores-Hampton 2011). However, such pathways are still highly neglected in Bangladesh.

The current chapter briefly discussed whether it is possible to meet national energy demand through biogas-based efficient energy generation technology, particularly for the overlooked, rural populace and to justify the efficiency as well as the economic cost, in relation to some grid-level renewable energy availability. Biogas technology influences the energy consumption and utilization by replacing various fuels and saving energy consumption. It produces renewable energy by using local input. It significantly benefits the environment in term of reduction of GHG emissions, and it benefits the agricultural practice. Biogas technology represents a sustainable way to produce energy for rural household, particularly in developing countries. More socio-technical studies need to be conducted for moving both science and policy forward in expanding biogas utilization across the economy for a more integrated consideration of energy and the environment. Without question, a multitude of renewable sources can be a versatile energy source for Bangladesh to mitigate the power crisis and while being economically competitive, as well as to properly dispose the waste, and to create employment of the local residence, For example, end-use applications such as space heating and transportation can be revitalized across economic sectors without much electrification, which is required for using solar and wind energy.

Further research on, not just the biogas technology, but the use-case details need to be conducted in order to realize the full potential of plant-based efficient energy generation system. If larger-scale developments are to be onboarded to ensure more gas and electricity production in rural settings, the institutional and technological considerations need to be adjusted significantly. The technical footprint may also need to be adjusted, to match the capability of the local workforce, and to better situate it near feedstock access and point of consumption. As a result, local innovators should plan to improve the design scheme of the biogas plant niche and regime to better fit into the existing landscape (Geels et al. 2017). This will not only make the energy source function better, but significantly reduce costs to build in comparison to other renewable energy plants in the rural context—ensuring reliable and clean fuel and electricity for those that need it the most. For instance, initial research finds that the feasibility of electricity production from biomass and biogas power plants can be further scaled. As an example, such proposed biogas power plants can be used to provide about 60–70% of total electricity in a small local area; with the rest coming from grid-tied integrations. More institutional and financial incentives need to be targeted to local communities in order to establish such plants in the remote village area, in a way that enables the production of enough electricity to power that particular village. Additionally, location plays a very important role in clean technology proliferation (Sunny and Shu 2017) as most plants need to be established near waste depot to minimize transportation losses and produce good amount of low-carbon electricity (Gomez 2013).

By providing a brief summary of the biofuel potential, we hope to contribute a socio-technical account of advancing multidisciplinary research, thereby elucidating numerous opportunities to address the low-carbon and highly resilient electricity issues in Bangladesh. Although the technical advancements in biogas technology is suboptimal, technical and agricultural professionals are constantly learning charting new trajectories of such energy use for multiple applications. At the same time, there are a few drawbacks that limit the reach of this technology and provide a case where other forms of renewable energy might be more appropriate. Biogas power plants need considerable geographic area but given such power plants need to be established within a small area, the locations on where they can be placed is rather limited. Often times, biogas plant may create noise and vibration but vibration, which requires additional attention from the plant's architects and civil engineers. Across its lifecycle, it may emit certain amounts of carbon dioxide gas as an externality, which varies across different contexts and need to be accounted for.

During the last 60 years, migration from rural-to-urban areas in Bangladesh increased alarmingly, from 2.5% people living in urban areas to 30%. Population growth itself rose from 0.35 million to more than 10 million in the capital of Dhaka. This has made the city the 8th mega city in the world, and according to the UN Population Division, putting it on target to be the 2nd mega city by 2015. It is because all energy supplies are limited to urban areas to cater to this increasing population. An appropriate mix of renewable energy sources as well as supporting institutional configurations is required in order to address the underlying causes of climate change.

References

Cheong D (2005) Studies of high rate anaerobic bio-conversion technology for energy production during treatment of high strength organic wastewaters

Geels FW, Sovacool BK, Schwanen T, Sorrell S (2017) Sociotechnical transitions for deep decarbonization. Science 357(6357):1242–1244

Ghimire PC (2005) Technical study of biogas plants installed in Bangladesh. Development Partners, Dhaka, pp 1–91

Gofran MA (2012) Biogas energy in Bangladesh. Ashraf Jahan Begum

Gomez CDC (2013). Biogas as an energy option: an overview. In: The biogas handbook, pp 1–16

IDCOL (2013) Biogas audit Bangladesh 2011–2013. NDBMP-IDCOL, Dhaka, Bangladesh

Khan EU, Mainali B, Martin A, Silveira S (2014) Techno-economic analysis of small scale biogas based polygeneration systems: Bangladesh case study. Sustain Energy Technol Assess 7:68–78

Launder K (2002) Energy crops and their potential development in Michigan. Michigan Biomass Energy Program

Mondal MAH, Kamp LM, Pachova NI (2010) Drivers, barriers, and strategies for implementation of renewable energy technologies in rural areas in Bangladesh—an innovation system analysis. Energy Policy 38(8):4626–4634

Nahar K (2011) Sweet Sorghum: an alternative feedstock for bioethanol. Iranica J Energy Environ 2(1):58–61

Nahar K (2012) Biogas production from water hyacinth (*Eichhornia crassipes*). Asian J Appl Sci Eng 1(1):9–13

Nahar K, Ozores-Hampton M (2011) Jatropha: an alternative substitute to fossil fuel. Horticultural Sciences Departments Florida: Institute of Food and Agriculture Science, University of Florida, pp 1–9

Nahar K, Sunny SA (2011) Extraction of biodiesel from a second-generation energy crop (*Jatropha curcas* L.) by transesterification process. J Environ Sci Technol 4:498–503

Nahar K, Sunny AS (2016) Biodiesel, glycerin and seed-cake production from roof-top gardening of *Jatropha curcas* L. Curr Environ Eng 3(1):18–31

Nahar K, Sunny SA, Shazi SS (2011). Land use requirement and urban growth implications for the production of biofuel in Bangladesh. Forest 1350(1350):0–92

Shiraiski K, Shirley R, Kammen DM, Huq S, Rahman F (2018) Identifying high priority clean energy investment opportunities for Bangladesh

Sunny S (2011) Green buildings, clean transport and the low carbon economy: towards Bangladesh's vision of a greener tomorrow

Sunny SA (2013) Globalization and complexity of environmental governance in sustainable development and climate change policy diffusion mechanisms in developing countries—the American response and the case of Bangladesh. J Sustain Dev Stud 3(2)

Sunny SA (2017) Systemic emergence under transitional uncertainty: the dynamic role of energy technology innovation. Kybernetes 46(9):1527–1541

Sunny SA, Shu C (2017) Investments, incentives, and innovation: geographical clustering dynamics as drivers of sustainable entrepreneurship. Small Bus Econ 1–23

Wikberg A, Blomberg M, Mathisen B (1998) Composition of waste from slaughterhouses, restaurants and food distributors. AFR-report (Sweden)

Zehnder AJ, Mitchell R (1978) Ecology of methane formation. Water Pollut Microbiol 2:349–376

Chapter 3
Climate Change Impact on Sundarbans: Challenges for Mitigation Strategies

Md. Mizanur Rahman, Md. Rakib Hossain, and Md. Nazrul Islam

Abstract The Sundarbans, the largest single block of tidal halophytic mangrove forest, has been facing increased challenges due to the combined effects of natural and anthropogenic disturbances. Climate change coupled with anthropogenic disturbances poses a great threat to the existence of this mangrove. Many regions of the world are affected by climate change, but Sundarbans is one of the highest affected regions due to high level of salinity, sedimentation, and land erosion. The salinity is increasing day by day due to frequent cyclones, sedimentation, and brackish tiger prawn cultivation. The increased salinity is jeopardizing the ecosystems of Sundarbans and poses more risk than any other stressors. The study aims to assess the impact of salinity on the pioneer and indicator plant species in terms of species distribution and the coping capacity with the increased salinity. Primary data was collected from 30 sample plots which were fresh swamp and fresh–brackish swamp in the past. Secondary historical data was collected from the Forest Department to understand the natural dynamics. It was found that the fresh swamp forests disappeared from their historical range. The pioneer species, *Heritiera fomes*, and the indicator species *Nipa fruticans* and *Phoenix pelludosa* are being replaced by the invasive species and highly salt-tolerant *Avicennia marina*. An immediate action is required to stop tiger prawn culture and to restore the government-owned canals from the encroachers.

Keywords Anthropogenic disturbances · Pioneer species · Indicator species · Salinity · Bangladesh

Md.Mizanur Rahman (✉) · Md.Rakib Hossain
Information and Communication Technology Division, Dhaka, Bangladesh

Md.Nazrul Islam
Department of Geography and Environment, Jahangirnagar University, Savar, Dhaka, Bangladesh
e-mail: nazrul_geo@juniv.edu

© Springer Nature Switzerland AG 2021   47
Md. Jakariya and Md. N. Islam (eds.), *Climate Change in Bangladesh*,
Springer Climate, https://doi.org/10.1007/978-3-030-75825-7_3

Introduction

The Sundarbans, the largest contiguous mangrove ecosystem in the world, is located in the southwest corner of Bangladesh, while the rest is in the West Bengal of India (Rahman and Alam 2020, Rahman and Vacik 2016). It consists of hundreds of mosaic islands webbed by canals, lagoons, tidal rivers, estuaries, and creeks (Raha et al. 2012). It was declared as the World Heritage Site by UNESCO in 1987, Global Biosphere Reserve in 1989, and the Ramsar Sites in 1992 (Rahman and Vacik 2015). The outflow of water from the Sundarbans delta is the third largest of the world. It is also transitional region between the freshwater originating from the Ganges and the saline water of the Bay of Bengal (Rahman and Vacik 2016). It is the sweet home of vascular plants, aquatic animals, birds, amphibians including crocodile, reptiles, mammals including the Bengal Tiger, crustaceans, algae, phytoplankton, zoo-plankton, and benthic invertebrates (Rahman and Vacik 2015). The Sundarbans is the habitat of 528 species of vascular plants (Rahman et al. 2015), 120 species of fishes, 35 species of reptiles, 270 species of birds, and 42 species of mammals (Rahman 2000). The ecosystems as well as the luxuriant biodiversity of Sundarbans have strong interactions with marine environments. The climatic stressors affecting Sundarbans are sea level rise, frequent cyclones, increased salinity, and changes in seasonal pattern. Near about 26% frequency of cyclones was increased over the last 120 years in the Bay of Bengal. Since 2006, Sundarbans experienced a series of cyclones including deadly *Sidr* and *Aila*. The Sundarbans are impacted by the cyclones by four ways: damage, tidal surge, siltation, and invasion by alien species (Rahman and Vacik 2016). Sundarbans has been experiencing both predictable and unpredictable climate change impacts since a long. The mangroves of Sundarbans are subject to multiple stressors originating from a rapidly changing environment and social dynamics (Alam et al. 2021, Mitchell et al. 2015). These stressors act simultaneously, degrading the diversity, ecosystem functions and services of the mangroves. Local pressures originate from the human, while (Rahman et al. 2007, 2009, 2010a, b; Rahman and Vacik 2009, 2010; Rahman 2009); global pressures originate from the climate change (Mitchell et al. 2015). The global problems exacerbate the local pressures (Burke et al. 2011; Gupta et al. 2007; Hoegh-Guldberg et al. 2007).

Many regions of the world are affected by climate change, but Sundarbans is one of the highest affected regions due to high level of salinity, sedimentation, and land erosion. Sea level rise is extremely prominent in Ganges basin, where Sundarbans delta is situated at the mouth of this basin. Any variation in the Ganges basin highly affects Sundarbans. In the last three decades, the rate of sea level rise was almost double of the global average. The plants are becoming shorter and narrower with fewer branches and leaves due to lower rates of photosynthesis. In 2010, the New Moore Island (South Talpatti) disappeared due to sea level rise couple oceanic erosion. It is predicted that many islands will be vanished from the continental shelf of the Bay of Bengal if sea levels keep the present pace.

Increased salinity raised red alert over degradation of water and soil, which is altering the ecosystems of Sundarbans (Rahman 2020, 2021a). The rise of salinity is intensifying the impacts of climate change. Local stressors like brackish shrimp culture and canal grabbing are worsening the scenarios.

With the change of mangrove ecosystems, the salinity is damaging the traditional livelihoods of the people residing in the fringe of Sunderbans (Rahman and Vacik 2014, Rahman 2021b, c). The increase in salinity is damaging the pioneer and indicator species,, and consequently, the timber stocks are depleting. The study aimed at examining the impacts of increased salinity on the abundance of pioneer and indicator plant species of Sundarbans mangrove and assessing the level of invasion by exotic plant species.

Methodology

A total number of 30 circular plots of 300 m^2 areas of each at the base of 30 canals along the eastern side of the *Passur* River were established (Rahman 2020). The mouth of the *Nandabala* Canal (the nearest canal from the human settlement) was the first canal and upwards canals to the south were taken as sample areas. Nearby water salinity, the density (number/plot) of different plant species and the percentage of affected trees by 'top dying' disease were examined at each plot. The secondary data over the last 20 years archived by the nearest forest offices was collected to assess the dynamics of the vegetations. The Pearson correlation coefficient was used to measure the strength of the linear relationship between the dependent and independent variables. Out of four ranges of Sundarbans, *Chandpai* Range was selected for data collection. Once, this range represented freshwater swamp and fresh–brackish water swamp. This range is famous for luxuriant biodiversity and covering 100,021 ha of forest lands. The forests and wetlands are very rich in plant and wildlife diversity. *Chandpai* is surrounded by the *Sarankhola* Range to the east and south, mainland to the north, and Khulna Range to the west. The study areas are comprised of a wildlife sanctuary, an ecologically critical area, a crocodile breeding center, a dolphin sanctuary, and two tourist spots. In addition, the vegetation types were observed in different beat (the smallest administrative unit) areas (Fig. 3.1).

Salinity Increase

Frequent tropical cyclones, namely Sidr, Nargis, Bijli, Aila, and Mahashen impacted Sundarbans through three primary mechanisms: wind damage, tidal surge, and sedimentation. Saline water is moving inwards due to consecutive cyclones in one decade. On the other hand, tidal inundation enlarges the areas of saline zone diminishing soil organic matter (Bazzaz et al. 1996) and reducing agricultural productivity (Afroz and Alam 2013). The increased salinity is jeopardizing the ecosystems of Sundarbans and poses more risk than any other stressors. It is becoming harder to mitigate salinity because of its long-lasting effects on the

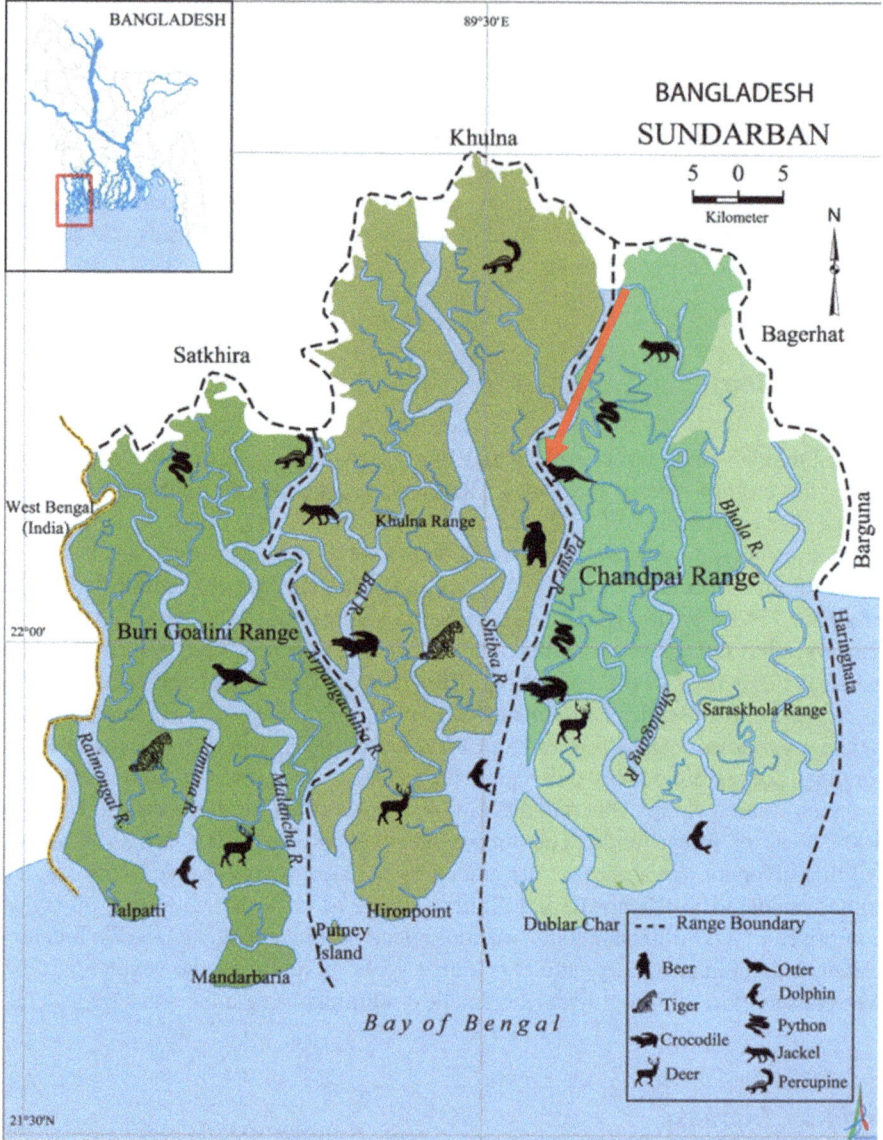

Fig. 3.1 Map showing four ranges of the Sundarbans (sample areas were marked by the red arrow) (Banglapedia 2020)

ecosystems. It affects the plants through concentrating in the root zones. Consequently, the previous freshwater rivers and canals have been converted into saline water bodies. Increased salinity has created a silent disaster in Rampal, Mongla, Morelganj, Shoronkhola, Dacope, Paikgachha, Koyra, Tala, Assasuni, and Sheyamnagar sub-districts. Saline water intrusion from the Bay of Bengal is exaggerated by tiger prawn culture and encroachment of the canals by land grabbers. The government-owned canals have become closed ended water bodies where the encroachers culture brackish shrimps. The blockades in the waterways of the canals resulted in lowest water volumes in the river, which bulled sedimentation and salinity levels. Brackish prawn culture plays a critical role in increasing salinity in the Sundarbans Delta (Bhowmick et al. 2016; Chowdhury and Maiti 2016; Rahman et al. 2013; Szabo et al. 2016). It increases the salinity level of river water, groundwater, and soil (Rahman et al. 2013; Dasgupta et al. 2014). Tiger prawn culture accelerates the depletion of soil nutrients and makes the adjacent and ground soils more acidic and saline which cannot be reclaimed (Ali 2006). The salinity is triple in the dry season comparing the monsoon (Mondal et al. 2001). The mechanism of salinization caused by tiger prawn cultivation and the consequences are shown (Fig. 3.2).

Once, there were five eco-regions in the Sundarbans: freshwater swamp, fresh–brackish swamp, brackish swamp (true mangrove), mangrove scrub, and littoral forest (Rahman 2020). The freshwater swamp forests of Sundarbans were characterized by slightly brackish water in the dry season, while quite fresh during monsoon. According to Champion and Seth (1968), the main plant species of this eco-region were and by *Sundari* (*Heritiera minor*), Mangrove Cannonball (*Xylocarpus molluccensis*), Black Mangrove (*Bruguiera conjugata*), *Keora* (*Sonneratia apetala*), Baen (*Avicennia officinalis*), Ora (*Sonneratia caseolaris*), Screw Pine (*Pandanus tectorius*), Chelwa (*Hibiscus tiliaceus*) and Nypa palm (*Nipa fruticans*). Fresh–brackish water swamp is an eco-region where freshwater

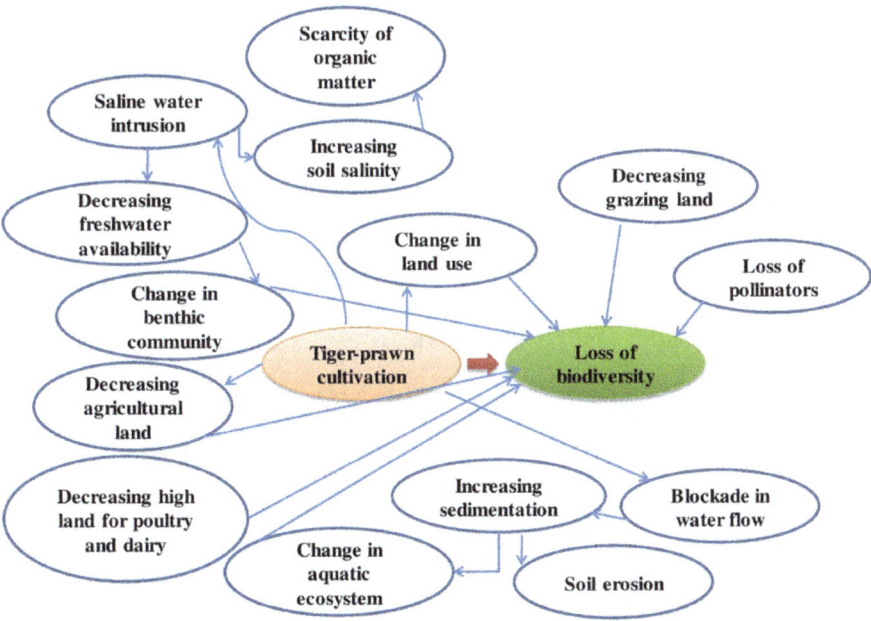

Fig. 3.2 Consequences of tiger prawn cultivation (Rahman 2020)

discharging from the river is mixed with the saline water. Sundari (*Heritiera fomes*), Keora (*Sonneratia apetala*), Ora (*S. caseolaris*), Passur (*Xylocarpus mekongensis*), Dhundul (*X. granatum*), Bain (*Avicennia alba*), Hantal (*Phoenix pelludosa*), Nypa palm (*Nipa fruticans*), swamp rice grass (*Leersia hexandra*), wild rice (*Potresia coarctata*), and Tiger fern (*Achrostichum aureum*) are the dominant plant species. In brackish swamp, Gewa (*Excoecaria agallocha*), Goran (*Ceriops decandra*), Keora (*Sonneratia apetala*), Sundari (*Heritiera minor*), and Nypa palm (*Nipa fruticans*) are the main plant species. In addition of the plant species of the brackish swamp, Jhau (*Tamarix indica*) is present in the littoral forest eco-region. Littoral forest is located along the shorelines of the Bay of Bengal. The mangrove scrub is highly affected saline eco-region which is inhabited by bushy shrubs. The study revealed that the fresh swamp eco-region totally disappeared from the Sundarbans. The characteristics and the vegetation composition of the fresh–brackish swamp eco-region are shifting to brackish swamp, where few portions of the brackish swamp have become mangrove scrub.

Rahman and Vacik (2015) reported that the pioneer species of the Sundarbans are suffering from increased salinity and the stocks are gradually declining. Thakur et al. (2012) reported that the Sundarbans is losing its own species due to increased salinity. Salinity affects natural regeneration and growth of the plants. Alteration of plant growth and succession are influenced by the salinity. With the increase of salinity, many plant species become dwarf, rare, and locally distinct. Most of the dominant plant species are affected by 'top dying' disease due to increased salinity (Rahman 1990). Salinity tolerance of various species of Sundarbans varies widely (Chaudhuri and Choudhury 1994; Naskar and Bakshi 1987; Waisel 1972), which determines the dominance, distribution, vegetation composition, structure, growth, and existence of species (Ball and Pidsley 1995; Chapman 1976; Das and Siddiqi 1985; Lin and Sternberg 1992; Portillo and Ezcurra 1989; Siddiqi et al. 1992; Waisel 1972). Salinity plays the most vital role in determining the mangrove vegetation patterns (Blasco 1977; Duke et al. 1998; Bhattacharjee et al. 2013; Ellison et al. 2000) and species distribution (Field 1995; Ellison et al. 2000; Duke et al. 1998; Bunt 1996; Banerjee et al. 2002) (Fig. 3.3).

Impact on Sundari (**Heritiera fomes***)*

The *Sundari* tree (*Heritiera fomes*) grows well in low saline condition. It was found that the salinity level had highly negative correlation with the water salinity level (Fig. 3.4).

The mangrove plant species are not slat lovers, but they are salt tolerant (Bowman 1917, Rahman 2020). Seed germination and growth of the plant are highly affected by increased salinity. Consequently, the plants become dwarf and disappear gradually. It was found that *Heritiera fomes* is being replaced by Keora (*Sonneratia apetala*). Mitra et al. (2004) found alteration in growth of mangroves due to salinity level. Salinity affects leaf structure, transpiration rate, conductance of

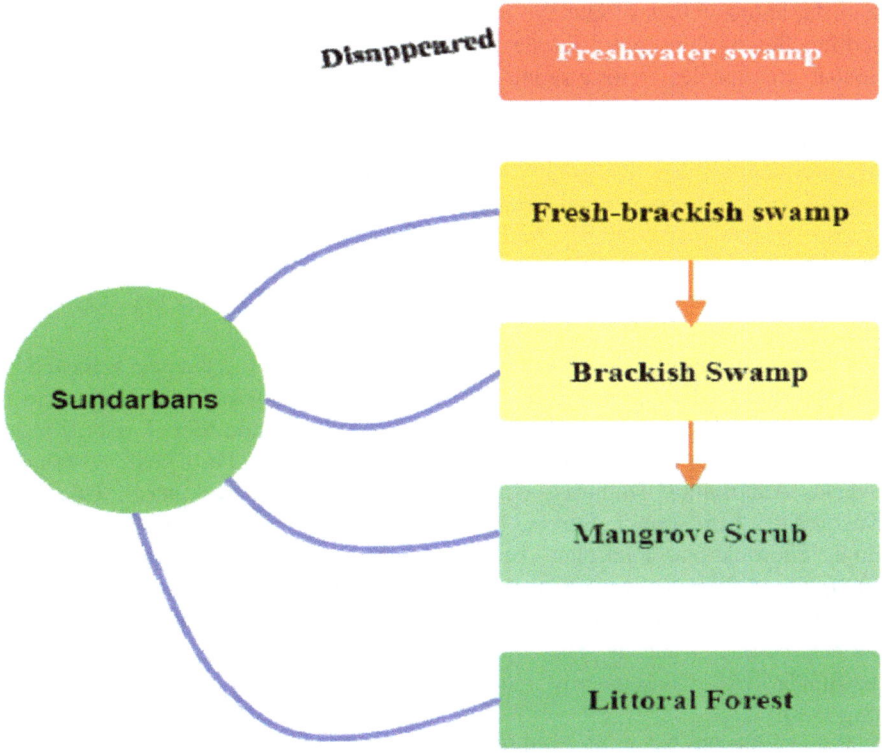

Fig. 3.3 Changes in the eco-regions of the Sundarbans (Rahman 2020)

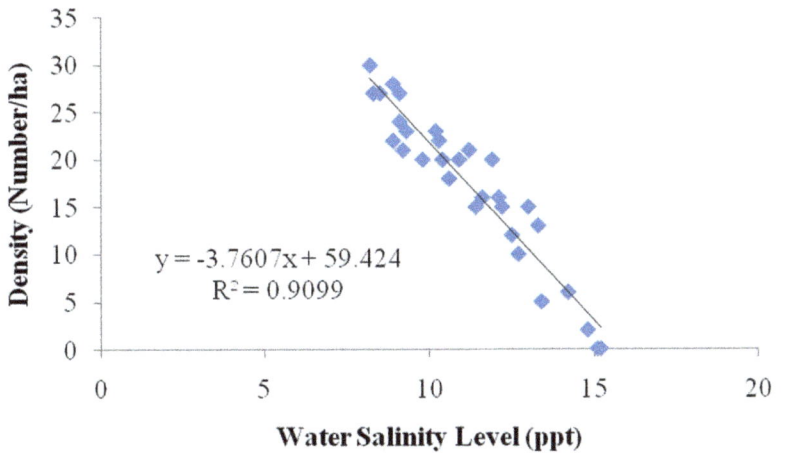

Fig. 3.4 Relationship between water salinity and density of the Sundari tree (Rahman 2020)

stomata, photosynthesis (Santiago et al. 2000; Parida et al. 2004), and the structure
and function of chloroplast (Parida et al. 2003). Mitra et al. (2004) revealed that the
Sundari tree species prefer a salinity range from 2 to 5 psu. Most of the mangrove
species in salinity level in between of 4 and 15 psu (Burchett et al. 1984; Clough
1984, 1985; Connor 1969; Downton 1982). Chaudhuri and Choudhury (1994)
reported that Sundari tree prefers lower salinity level. The tree responds physio-
logically and shows abnormal behaviors at 15 psu. The seedlings cannot adapt at
20 psu. Soil salinity regulates the distribution and abundance of *Heritiera fomes*.
The infestation of 'top dying' disease has been intensified in the recent time. It is
found that salinity level has strong positive relationship with the 'top dying'
disease.

In addition, through the observational study and analyzing the data achieved by
different forest offices at bit level, it was found that *Heritiera fomes* declined from
Buddhomari, Mrigamari, Dangmari, Karamjal, Andharmanik, Mora Passur, Jongra,
Nandobala, Harbouria, Jeodhara, Choraputia, Katakhali, Pashakhali, Boroitola,
Shapla, Amurbunia, Dumuria, Gulishakhali, Dhansagor, Kolomtezi, Chandpai,
Kochikhali, Tamulbunia, Supoti, Mora Bogi, Nangli, Charkhali, Chandeshwar,
Bogi, Shoronkhola, Panirghat, Bhola, and Dashervarani forest areas (Rahman
2020). Once, these areas were inhabited by the dominant *Sundari* species. It can be
projected that the Sundari tree species will be disappeared from the Sundarbans in
the next couple of decade. The nomenclature of the Sundarbans was derived from
the Sundari tree. Consequently, the Sundarbans will lose its current name as well
(Fig. 3.5).

Rahman (1990) found 45 million affected trees by top dying disease which was
about 20% of the entire. The scenario drastically declined over the last two decades.
Now, the infestation rate ranges from 40 to 82%. Islam and Wahab (2005) reported

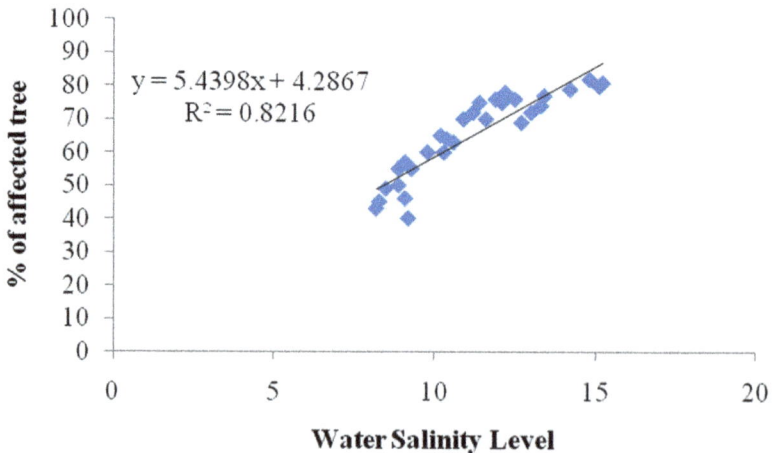

Fig. 3.5 Relationship between the percentages of affected Sundari tree by 'top dying' disease and
water salinity level (Rahman 2020)

Fig. 3.6 Affected Sundari trees by 'top dying' disease

that approximately 70% of *H. fomes* stems were affected by 'top dying' disease. Thakur et al. (2012) reported that top dying disease is caused by multiple factors like increased salinity, reduced water flow, irregular periodic inundation, sedimentation, and cyclone-induced stress. Rahman (1990) revealed that older trees are more vulnerable to 'top dying' disease. Increased increases in osmotic stress on the roots cause unavailability of water and nutrients. 'Top dying' causes death and truncated of the younger tree (Rahman 1994) (Fig. 3.6).

Impact on **Keora**

Mangrove apple (*Sonneratia apetala*) is locally known as Keora. This species is considered one of the pioneer species of Sundarbans (Rahman 2020, Nasrin et al. 2017). In the newly accreted land, it appears first and stabilizes the soil and creates congenial conditions for other species. Thereby, it prevents soil erosion caused by tidal surges and wave energy (Rahman 2020). For this reason, it is used for reclamation of degraded mangroves. This species can cope with siltation or sedimentation and anaerobic substrate and can exploit nutrients from different sources. The flowers consists large amount of nectars which attract the pollinators easily.

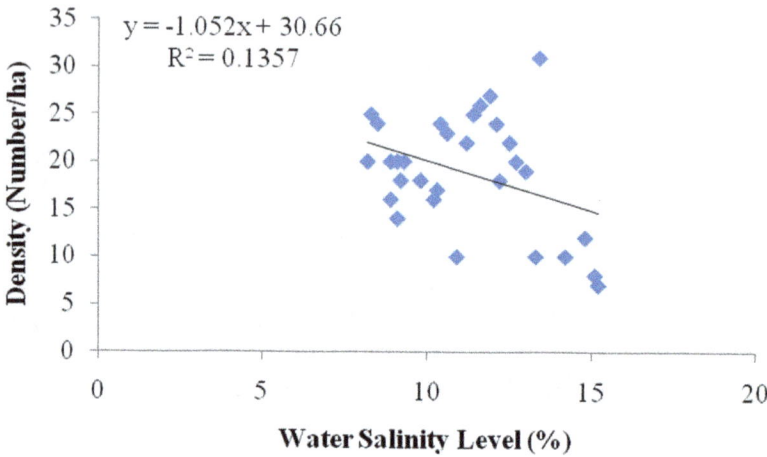

Fig. 3.7 Relationship between the density of Keora and water salinity level (Rahman 2020)

It is highly resilient in any disaster as any physical damage is repaired within the shortest time (Fig. 3.7).

The study found that there was slightly negative relationship between the density of Keora and water salinity level. Hoque et al. (1999) found that seed germination of Keora was affected by increased salinity. Rahman (2020) reported that Keora is distributed in the wide range of salinity and can adapt at best 20 ppt of salinity level, but the dispersal and behavior are regulated by the salinity level (Hema and Ghose 2003). *Keora* plays dual as in salt regulation; it can exclude and accumulate salt simultaneously (Hutchings and Saenger 1987; Saenger 1982). The study found that there was slightly negative relationship between the density of *Keora* and water salinity level.

Impact on Mangrove Date Palm (Phoenix paludosa)

Mangrove date palm (*Phoenix paludosa*) is locally known as *Henthal*. This is one of the dominant undergrowth species of Sundarbans (Rashid et al. 2008), which forms small clumps in mosaic pattern especially at edge of the or in the interior parts of Sundarbans. Healthy palms are available in the long strips at the edges of forest locating on the river bank (Hossain 2003).

Figure 3.8 shows a strong negative relationship between the density of the mangrove date palm and water salinity level. Rashid et al. (2008) reported that mangrove date palm cannot tolerate frequent inundation and strong salinity. The salinity also determines the growth and health conditions as well.

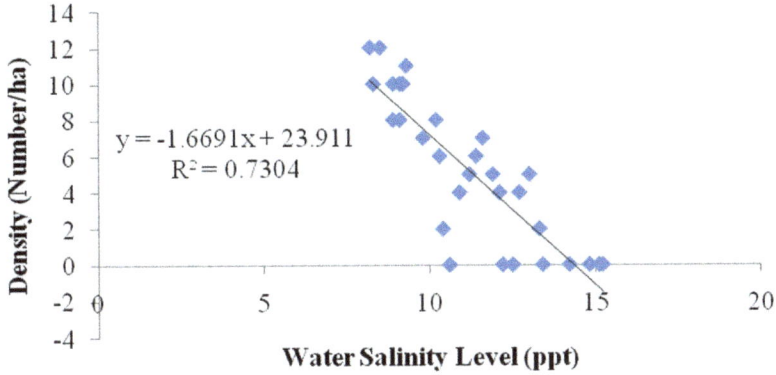

Fig. 3.8 Relationship between the density of mangrove date palm and water salinity level (Rahman 2020)

Impact on Nypa *Palm* (Nipa fruticans)

Nypa palm (*Nypa fruticans*) is one of the pioneers as well as the indicator species of Sundarbans Manrgove. Locally, it is known as *Golpata*. Nypa is a useful and fairly common species of mangrove forests of Asia and Oceania (Hamilton and Murphy 1988). Nypa can grow in a wide range of habitats from inland to seawater habitat (Theerawitaya et al. 2014). It is neither a true littoral nor a high saline resistant species. *Nypa* grows well in the mudflat and riparian habitats where the water rises and falls back slowly. It can tolerate irregular inundation until as the soil dries out.

The study revealed that there was highly negative relationship between the density of *Nypa* palm and water salinity level. Theerawitaya et al. (2014) reported that *Nypa* palm grows well in the moderate saline zone. In another studies, Rashid et al. (2008) and Rahman (2020) revealed that *Nypa* palm prefers low salinity (Fig. 3.9).

Impact on Morcha Baen (Avicennia marina)

Gray mangrove or white mangrove *(Avicennia marina) is locally known as Morcha Baen. These species are common in true mangrove or brackish swamp* (Barik and Chowdhury 2014).

Figure 3.10 indicates that there was a strong positive negative correlation between the salinity level and the density of white mangrove. The historical data collected from the forest offices revealed that there was no white mangrove in the study area 30 years back. The species can tolerate 1.5 times (Martin et al. 2010) or twice of salinity existing in the sea water (Reef et al. 2010). In addition, this species can tolerate acidity, high temperature, and frost (Morrisey et al. 2010). This species

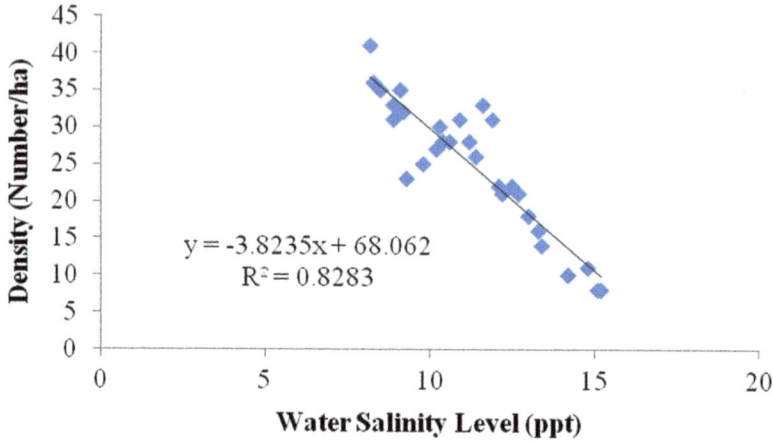

Fig. 3.9 Relationship between the density of Nypa palm and water salinity level (Rahman 2020)

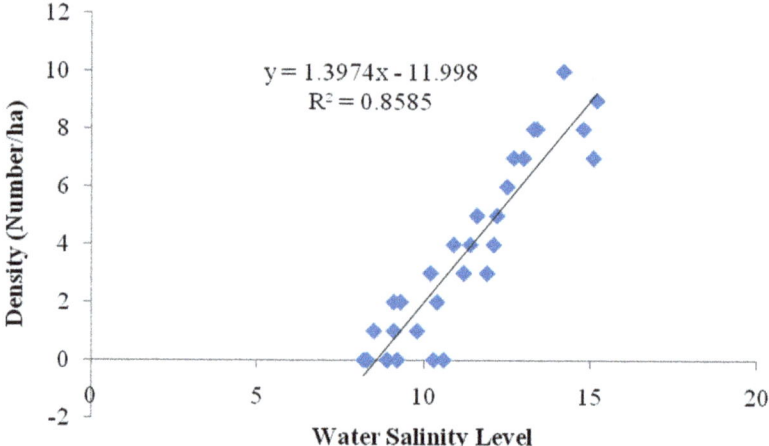

Fig. 3.10 Relationship between the density of white mangrove and water salinity level (Rahman 2020)

can adapt in the habitats having salinity from 20.9 to 31.25 ppt (Hema and Ghose 2003).

Invasion

Biological invasions are treated as one of the greater threats to the biodiversity of Sundarbans. The Sundarbans is highly degraded by high levels of habitat destruction and fragmentation creating small fragmented habitat mosaics. The microclimatic conditions facilitate favorable condition for the establishment of opportunist invasive species. Native plant communities are susceptible to invasion. Anthropogenic disturbances and frequent cyclones cause clear felling, toppling, and uprooting of the pants. As a result, small fragmentations and gaps are created, which invite the invaders. Habitat degradation and alteration of vegetation are occurring simultaneously. They compete with the native species for water, light, and nutrients; hamper natural regeneration of the native species; cause physical damage, degrade water quality; and alter habitat. Habitat modification and the invasion cause decline and even extinction of the native species.

A total number of 25 invasive species including 06 species of trees, 06 species of climbers, 04 species of shrubs, 03 species of herbs, 02 species of grasses, 02 species of epiphytes, and 01 species of ferns were found in the study area (Rahman 2020). The highest abundant species were Common Derris (*Derris trifoliate*) followed by common water hyacinth (*Eichhornia crassipes*) and Siam weed (*Eupetorium odoratum*). Biswas et al. (2007) found 23 invasive species in the Sundarbans. In another study, Rahman and Vacik (2016) reported 23 invasive species followed by tropical cyclone *Aila*. Biswas et al. (2007) identified common Derris (*Derris trifoliate*), common water hyacinth (*Eichhornia crassipes*), and Siam weed (*Eupetorium odoratum*) as the most invasive species.

Recommendations

Bangladesh has nothing to do in managing global stressors, but can control the local pressures originating from the anthropogenic disturbances. Bangladesh should respond immediately against the local problems causing increased salinity.

Protecting the Canals

The state-owned canals in the Sundarbans delta should be restored and protected from the land grabbers and encroachers. This is immensely important to increase the water flows of the rivers to control sedimentation and salinity level. It is also important for restarting agricultural practices. The obstructions on the canals' natural water flow should be cleared immediately.

Stopping Tiger Prawn Cultivation

Salinization in surface water due to tiger prawn cultivation and saline water intrusion due to frequent cyclones caused a serious environmental and

Table 1 The list of invasive plant species (Rahman 2020)

Common name	Scientific name	Plant types
Mikania scandens	Climbing hempweed	Climber
Derris trifoliata	Common Derris	Climber
Flagellaria indica	Hell tail	Climber
Sarcolobus globosus Wall	Pitcher plant	Climber
Eupatorium odoratum L.	Siam weed	Climber
Entada rheedii	Snuff box sea bean	Climber
Dendrophthoe falcata (L.f.) Etting	Honey Suckle Mistletoe	Epiphyte
Hoya parasitica (Roxb.) Wall. ex Wight	Red Hoya	Epiphyte
Acrostichum aureum Linn	Golden Leather Fern	Fern
Arundo donax L.	Giant reed	Grass
Imperata cylindrica (L.) Raeuschel	Speargrass	Grass
Saccharum spontaneum	Wild sugarcane	Grass
Eichhornia crassipes	Common water hyacinth	Herb
Lantana camara Linn	Lantana	Herb
Typha angustata Borry f	Lesser bulrush	Herb
Clerodendrum inerme (L.) Gaertn	Garden quinine	Shrub
Cryptocóryne ciliáta (Roxb.) Fisch. ex	Water trumpet	Shrub
Hibiscus tiliaceus	Sea rosemallow	Shrub
Ipomoea fistulosa	Bush morning glory	Shrub
Vachellia nilotica	Gum arabic tree	Tree
Pongamia pinnata (L.) Pierre	Indian beech	Tree
Syzygium fruticosum (Roxb.) DC	Java plum	Tree
Salacia chinensis	Lolly berry	Tree
Excoecaria indica (Wild.) Muell.-Arg	Mock-willow	Tree
Tamarix indica L.	Saltcedar	Tree

socioeconomic degradation in the Sundarbans delta. Increased salinity is being treated as a silent disaster in that region. Brackish shrimp culture caused massive loss of field crop, horticultural crops, and dairy–poultry industries. Fresh drinking water crises are at the highest level. An integrated income-generating activities have been prescribed here to overcome these challenges (Fig. 3.11).

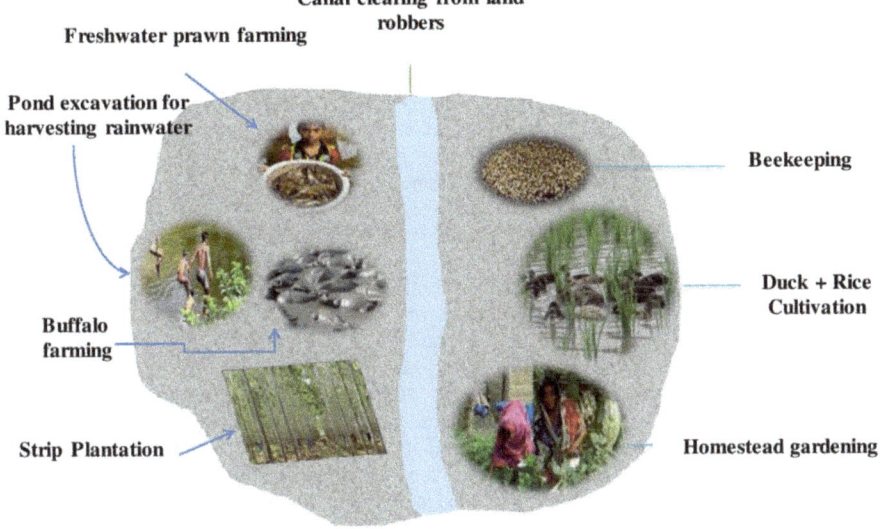

Fig. 3.11 Integrated farming systems alternative to brackish shrimp culture (Rahman 2020)

References

Afroz T, Alam S (2013) Sustainable shrimp farming in Bangladesh: a quest for an integrated coastal zone management. Ocean Coast Manage J 71:275–283

Alam S, Rahman MM, Arif AA (2021) Challenges and opportunities in artisanal fisheries (Sonadia Island, Bangladesh): the role of legislative, policy and institutional frameworks. Ocean Coastal Manage 201:105424. https://doi.org/10.1016/j.ocecoaman.2020.105424

Ali AM (2006) Rice to shrimp: land use land cover changes and soil degradation in Southwestern Bangladesh. Land Use Policy 23(4):421–435

Ball MC, Pidsley SM (1995) Growth responses to salinity in relation to distribution in two mangrove species, *Sonneratia alba* and *Sonneratia lanceolata*, in northern Australia. Funct Ecol 9:77–85

Banerjee LK, Rao TA, Shastry ARK, Ghosh D (2002) 'Diversity of coastal plant communities in India', ENVIS-EMCBTAP. Botanical Survey of India, Ministry of Environment and Forests, Kolkata

Banglapedia (2020) Map of the Sundarbans. Encyclopedia of Bangladesh

Barik J, Chowdhury S (2014) True mangrove species of sundarbans delta, West Bengal, Eastern India. Check List 10(2):329–334

Bazzaz FA, Sombroek WG, Food and Agriculture Organization of the United Nations (1996) Global climate change and agricultural production: direct and indirect effects of changing hydrological, pedological, and plant physiological processes. Food and Agriculture Organization of the United Nations (FAO), Rome

Bhattacharjee AK, Zaman S, Raha AK, Gadi SD, Mitra A (2013) Impact of salinity on above ground biomass and stored carbon in a common mangrove *Excoecaria agallocha* of Indian Sundarbans. Am J Bio Pharmacol Biochem Life Sci 02:1–11

Bhowmick B, Uddin Z, Rahman S (2016) Salinity changes in south west Bangladesh and its impact on rural livelihoods. Bangladesh J Vet Med 14(2):251–255

Biswas SR, Choudhury JK, Nishat A, Rahman MM (2007) Do invasive plants threaten the Sundarbans mangrove forest of Bangladesh? For Ecol Manage 245:1–9

Blasco F (1977) Outline of ecology, botany and forestry of the mangals of the Indian subcontinent, West Coastal Ecosystems'. Elsevier Scientific Publishing Company, Oxford

Bowman (1917) Mangrove regeneration and management. A.K.F. Hoque, 1995, Mimeograph

Bunt JS (1996) Mangrove zonation: an examination of data from seventeen riverine estuaries in tropical Australia. Ann Bot 78:333–341

Burchett MD, Field CD, Pulkownik A (1984) Salinity, growth and root respiration in the grey mangrove *Avicennia marina*. Physiol Plant 60:113–118

Burke L, Reytar K, Spalding M, Perry A (2011) 'Reefs at risk revisited', Report. World Resource Institute, Washington, DC

Chapman VJ (1976) Mangrove vegetation. J. Cramer, Vaduz, Liechtenstein

Chaudhuri AB, Choudhury A (1994) Mangroves of the sundarbans, India, vol I. IUCN, p 284

Chowdhury A, Maiti SK (2016) Identifying the source and accessing the spatial variations, contamination status, conservation threats of heavy metal pollution in the river waters of Sundarban biosphere reserve. J Coast Conserv 20(3):257–269

Clough BF (1984) Growth and salt balance of the mangroves *Avicennia marina* (Forsk.) Vierh, and *Rhizophora slylosa* griff in relation to salinity. Aust J Plant Physiol 11:419–430

Clough BF (1985) Effect of nutrient supply on photosynthesis in mangroves In: Bhosale LJ (eds) The mangroves. Proceedings of national symposium biology utilisation and conservation mangroves. Shivaji University, Kohlapur, India, pp 80–88

Connor DJ (1969) Growth of grey mangrove (*Avicennia marina*) in nutrient culture. Biotropica 1:36–40

Das S, Siddiqi NA (1985) The mangroves and mangrove forests of Bangladesh, Mangrove Silviculculture Division, Bulletin No. 2, Bangladesh Forest Research Institute, p 142

Dasgupta S, Kamal FA, Khan ZH, Choudhury S, Nishat A (2014) River salinity and climate change: evidence from coastal Bangladesh. The World Bank

Downton WJS (1982) Growth and osmotic relations of the mangrove *Avicennia marina*, as influenced by salinity. Aust J Plant Physiol 9:519–528

Duke NC, Ball MC, Ellison JC (1998) Factors influencing biodiversity and distributional gradients in mangroves. Glob Ecol Biogeogr Lett 7(1):27–47

Ellison AM, Mukherjee BB, Karim A (2000) Testing patterns of zonation in mangroves: scale dependence and environmental correlates in the Sundarbans of Bangladesh. J Ecol 88:813–824

Field CD (1995) Impact of expected climate change on mangroves. Hydrobiologia 295:75–91

Gupta J, van der Leeuw K, de Moel H (2007) Climate change: a 'glocal' problem requiring 'glocal' action. Environ Sci 4:139–148

Hamilton LS, Murphy DH (1988) Use and management of Nipa palm (*Nypa fruticans*, arecaceae): a review. Econ Bot 42:206–213

Hema J, Ghose M (2003) Forest strucure and species distribution along soil salinity and pH gradient in mangrove swamps of the sunderbans. J Trop Ecol 44(2):197–206

Hoegh-Guldberg O, Mumby PJ, Hooten AJ, Steneck RS, Greenfield P, Gomez E et al (2007) Coral reefs under rapid climate change and ocean acidification. Science 318:1737–1742

Hoque AKM, Kabir ME, Islam MS (1999) Effect of salinity on the germination of *Sonneratia apetala* Buch.-Ham, Bangladesh. J For Sci 28(1):32–37

Hossain ABME (2003) The undergrowth species of Sundarban mangrove forest ecosystem (Bangladesh). The final report on Sundarban Biodiversity Conservation Project, IUCN, Dhaka, Bangladesh

Hutchings P, Saenger P (1987) Ecology of mangroves. University of Queensland Press, Australia

Islam MS, Wahab MA (2005) A review on the present status and management of mangrove wetland habitat resources in Bangladesh with emphasis on mangrove fisheries and aquaculture. Hydrobiologia 542:165–190

Lin GH, Sternberg LSL (1992) Effect of growth form, salinity, nutrient and sulfide on photosynthesis, carbon isotope discrimination and growth of red mangrove (*Rhizophora mangle* L.). Funct Plant Biol 19(5):509–517

Martin PH, Canham CD, Kobe RK (2010) Divergence from the growth-survival trade-off and extreme high growth rates drive patterns of exotic tree invasions in closed-canopy forests. J Ecol 98:778–789

Mitchell SB, Jennerjahn TC, Vizzini S, Zhang W (2015) Changes to processes in estuaries and coastal waters due to intense multiple pressures–an introduction and synthesis. Estuar Coast Shelf Sci 156:1–6

Mitra A, Banerjee K, Bhattacharyya DP (2004) The other face of mangroves, Department of Environment, Govt. of West Bengal, India

Mondal MK, Bhuiyan SI, Franco DT (2001) Soil salinity reduction and prediction of salt dynamics in the coastal rice lands of Bangladesh. Agric Water Manage 47:9–23

Morrisey DJ, Swales A, Dittmann S, Morrison MA, Lovelock CE, Beard CM (2010) The ecology and management of temperate mangroves. Oceanogr Mar Biol Annu Rev 48:43–160

Naskar KR, Bakshi DNG (1987) Mangrove swamps of the sundarbans—an ecological perspective. Naya Prokash 1–263

Nasrin S, Hossain M, Alam MR (2017) A monograph on *Sonneratia apetala* Buch.-Ham. Lambert Academic Publishing

Parida AK, Das AB, Mittra B (2003) Effects of NaCl stress on the structure, pigment complex composition and photosynthetic activity of mangrove *Bruguiera parviflora* chloroplasts. Photosynthetica 41:191–200

Parida AK, Das AB, Mittra B (2004) Effects of salt on growth, ion accumulation, photosynthesis and leaf anatomy of the mangrove *Bruguiera parviflora*. Trees Struct Funct 18:167–174

Portillo JL, Ezcurra E (1989) Response of three mangroves to salinity in two geoforms. Funct Ecol 3(3):355–361

Raha A, Das S, Banerjee K, Mitra A (2012) Climate change impacts on Indian sunderbans: a time series analysis (1924–2008). Biodivers Conserv 21:1289–1307

Rahman MA (1990) A comprehensive report on Sundri (*Heritiera fomes*) trees with particular reference to top dying in the sundarbans. In: Rahman MA, Khandakar KFU, Ahmed Ali MO (eds) Proceedings of the seminar on top dying of Sundri (*Heritiera fomes*) trees, Bangladesh Agricultural Research Council, Farmgate, Dhaka, Bangladesh, pp 12–63

Rahman MA (1994) Diseases and wood decay of tree species with particular reference to top dying of sundri and the magnitude of its damage in the Sundarbans in Bangladesh. In: Proceedings of the national seminar on integrated management of ganges flood plains and sundarbans ecosystem, Khulna University, Bangladesh, pp 35–39

Rahman LM (2000) The sundarbans: a unique wilderness of the world. USDA Serv Proc 2:143–148

Rahman MM (2009) Plant diversity and anthropogenic disturbances in the Sal (*Shorea robusta* C. F. Gaertn) forests of Bangladesh. PhD thesis, University of Natural Resources and Life Sciences, Vienna, Austria

Rahman MM (2020) Impact of increased salinity on the plant community of the Sundarbans Mangrove of Bangladesh. Commun Eco 21:273–284

Rahman MM (2021a). Achieving agenda 2030 in Bangladesh: the crossroad of the governance and performance. Public Administration and Policy. https://doi.org/10.1108/PAP-12-2020-0056

Rahman MM (2021b) Can ordinary people seek environmental justice in Bangladesh? Bangladesh J Public Admin 29(2):15–34

Rahman MM (2021c) Assessing the progress and pitfalls of the Ministry of environment, forest, and climate change in achieving SDGs in Bangladesh. Bangladesh J Public Admin 29(2):140–158

Rahman MM, Alam A (2020) Regulatory and institutional framework for the conservation of coral reefs in Bangladesh: a critical review. In: Alam MA, Alam F, Begum D (eds) Knowledge management, governance and sustainable development: lessons and insights from developing countries, Rutledge, India, pp 231–244. https://doi.org/10.2139/ssrn.3794550

Rahman MM, Vacik H (2009) Can picnic influence floral diversity and vitality of trees in Bhawal National Park of Bangladesh? For Stud Chin 11(3):148–157

Rahman MM, Vacik H (2010) Vegetation analysis and tree population structure of Sal (*Shorea robusta*) forests: a case study from the Madhupur and Bhawal National Park in Bangladesh. In: Polisciano G, Farina O (eds) National parks: vegetation, wildlife and threats. Nova Science Publishers, New York, USA

Rahman MM, Vacik H (2014) Impact of climate change on the nipa palm of sundarbans (Parrotta JA, Moser CF, Scherzer AJ, Koerth NE, Lederle DR (eds) Sustaining forests, sustaining people: the role of research, XXIV IUFRO World Congress, 5–11 October 2014, Salt Lake City, USA) Int For Rev 16(5)

Rahman MM, Vacik H (2015) Response of pioneer mangrove tree species of the sundarbans to increased salinity. In: AGU Chapman conference on "the width of the tropics: climate variations and their impacts", 27–31 July 2015, Sante Fe, USA

Rahman MM, Vacik H (2016) Recruitment of invasive plant species in the Sundarbans following tropical Cyclone Aila. In: American Geophysical Union, Ocean Sciences Meeting 2016, New Orleans, USA

Rahman MM, Begum F, Nishat A, Islam KK, Ruprecht H, Vacik H (2007) Comparison of structural diversity of tree-crop associations in peripheral and buffer zones of Gachabari Sal forest area, Bangladesh. J for Res 18(1):23–26

Rahman MM, Nishat A, Vacik H (2009) Anthropogenic disturbances and plant diversity of the Madhupur Sal forests (*Shorea robusta* C.F. Gaertn) of Bangladesh. Int J Biodivers Sci Manag 5(3):162–173

Rahman MM, Begum F, Nishat A, Islam KK, Vacik H (2010a) Species richness of climbers in natural and successional stands of Madhupur Sal (*Shorea robusta* c.f. gaertn) forest, Bangladesh. J Trop Subtropical Agroecosyst 12:117–122

Rahman MM, Rahman MM, Islam KS (2010b) The causes of deterioration of Sundarban mangrove forest ecosystem of Bangladesh: conservation and sustainable management issues. AACL Bioflux 3(2):77–90

Rahman M, Giedraitis VR, Lieberman LS, Akhtar T, Taminskienė V (2013) Shrimp cultivation with water salinity in Bangladesh: the implications of an ecological model. Univ J Public Health 1(3):131–142

Rashid SH, Böcker R, Hossain ABME, Khan SA (2008) Undergrowth species diversity of Sundarban mangrove forest (Bangladesh) in relation to salinity Ber. Inst. Landschafts-Pflanzenökologie Univ. Hohenheim Heft 17:41–56

Reef R, Feller IC, Lovelock CE (2010) Nutrition in mangroves. Tree Physiol 30:1148–1160

Saenger P (1982) Morphological, anatomical and reproductive adaptations of Australian mangroves. In: Clough BF (ed) Mangrove ecosystems in Australia: structure, function and management. Australian National University Press, Canberra, Australia, pp 153–191

Santiago LS, Lau TSP, Melcher J, Steele OC, Goldstein G (2000) Morphological and physiological responses of Hawaiian *Hibiscus tiliaceus* population to light and salinity. Int J Plant Sci 161:99–106

Siddiqi NA, Khan MAS, Islam MR, Hoque AKF (1992) Underplanting-a means to ensure sustainable mangrove plantations in Bangladesh. Bangladesh J for Sci 21:1–6

Szabo S, Hossain MS, Adger WN, Matthews Z, Ahmed S, Lázár AN, Ahmad S (2016) Soil salinity, household wealth and food insecurity in tropical deltas: evidence from south-west coast of Bangladesh. Sustain Sci 11(3):411–421

Thakur AK, Behera MD, Navania N, Roy P, Vikas P, Mathur RP, Singh RP, Ramasre P, Roorkee Pathania R, Behera S, Bora U, Tare V (2012) The status of Sundari (*H. fomes*) an indicators species in the Sunderbans: the lower Ganga River Basin, Report, Indian Institutes of Technology

Theerawitaya C, Samphumphuang T, Cha-Um S, Yamada N, Takabe T (2014) Responses of Nipa palm (*Nypa fruticans*) seedlings, a mangrove species, to salt stress in pot culture. Flora 209:597–603

Waisel Y (1972) Biology of halophytes. Academic Press, New York and London

Chapter 4
Climate Change and Sustainability of Agriculture in Bangladesh

Nazmul Ahsan Khan

Abstract Bangladesh is forced to import food from the international market or resort to foreign aid. Droughts or cyclones both disrupt agricultural output rapidly and thus leave a lot of families helpless. They also affect domestic food prices which makes it difficult for the majority of the middle-income families within the urban areas. In Bangladesh, food security has been one of the major national priorities for last few decades, but the target has always been interrupted by the climate change and for resource constraints. Present section of this chapter will highlight the major effects of climate change in the food production and the national resources constraints to address the food security. However, major constraints in terms of food security in Bangladesh attributed to cultivable land scarcity, irrigation water scarcity in summer, lack of technological knowledge, lack of climate adaptive crop variety, lack of institutions and professionals as well as social and cultural constraints. Richer farmers can afford modern machineries, genetically modified crop seeds and chemical fertilizers. This results in efficient farming, higher yield from a unit plot of land or better utilization of larger farmlands. This not only produces good quality and large quantity of output, but also means that the produces can be sold at cheaper rates at the local market, or can be processed and exported for higher rates. Farmers who cannot afford such technology are at a disadvantage.

Keywords Climate change · Food security · Crop variety · Sustainable agriculture · Bangladesh

N. A. Khan (✉)
Environmental Science and Management, North South University, Bashundhara R/A, Dhaka 1229, Bangladesh
e-mail: nazmul.khan@northsouth.edu

© Springer Nature Switzerland AG 2021
Md. Jakariya and Md. N. Islam (eds.), *Climate Change in Bangladesh*,
Springer Climate, https://doi.org/10.1007/978-3-030-75825-7_4

Introduction

Bangladesh, primarily being a low-lying country with an intricate network of rivers, benefits from having alluvium-rich soil are good for farming. As such, communities grew around the major river networks of the Ganges, Brahmaputra and Meghna, and their associated tributaries. These rivers have fertile floodplains which allow for high crop yield. A lot of subsistence-based farmers depend on the river network for their day-to-day food source. This developed an almost pendulum liked back and forth interaction with the rural community and nature. River networks frequently cause flooding which disrupts livelihood, but the receding flood leaves behind silt and alluvium-rich soil which in turn increases productivity for the next season's harvest. The benefits seem to outweigh the risks which is why so many rural communities cluster around the rivers.

The importance of agriculture is evident from its contributions to both Bangladesh's GDP and employment level. Almost 20% of the total GDP stems from agriculture while simultaneously employing 60% of the population. Furthermore, around 60% of the total landmass is also considered to be arable.

Agricultural produces in Bangladesh range from rice, which is primarily for domestic consumption, to jute and tea, which are the main export crops. In addition to these three main crops, farmers also produce sugarcane, tobacco, cotton, various fruits such as jackfruit, banana, potatoes, pineapples, etc. for local consumption. Since rice is a staple food for Bangladesh, they are produced in plenty. Both rice and wheat production plays a crucial role in achieving self-sufficiency, but unpredictable weather condition causes occasional interference. As such, Bangladesh is forced to import food from the international market or resort to foreign aid. Droughts or cyclones both disrupt agricultural output rapidly and thus leave a lot of families helpless. They also affect domestic food prices which makes it difficult for the majority of the middle-income families within the urban areas.

While our ancestors relied on hunting and gathering during the dawn of time, it was only sufficient to support a small group of population. They always had to be on the move in order to not exhaust the supplies of a particular area. As a result, it was difficult to have a thriving population as food security was lacking. However, once they realized that it is far more efficient to settle down in one particular location and grow their own food, they suddenly began to have a more stable life. The advent of agriculture meant that a population could finally settle down and not worry about supplies as much as before. This caused the start of early civilization around river networks that had fertile soil near its floodplains.

As Malthusian theory have stated, population grows exponentially while agricultural outputs expand arithmetically. This was valid during its time, where technological innovation could not feasibly be taken into account. However, with industrial revolution and its subsequent green revolution, the prediction no longer stayed true. Technological innovation meant that some of the labor involved in food production could be replaced by machineries, and with the same amount of land, more yield could be produced with the adoption of intensive practices.

For low-income countries, such as Bangladesh itself, the advent of green revolution marked the time where reliance on imported food could be lessened and
even local produce could outpace local consumption. This meant that agricultural
products could, for the very first time, be exported and act as a source of foreign
revenue.

However, the effects of the green revolution did not manifest itself equally across
a countries geography. The technology and synthetic nature of the revolutions
output meant that it only benefited the richer segment of the producers. Small,
subsistence-based farmers were often deprived of the benefits and instead were at a
competitive disadvantage as they could not keep up with the high initial costs of
production. This inequality meant that a large portion of the farming community
were excluded from the benefits and had to rely on a poor resource base.
Furthermore, the intensive application of synthetic practices meant that nutrients
depleted rapidly and caused ecological deterioration and harm. While the overall
output of the country has increased, the green revolution widened inequality and
brought its output at the expense of environmental stability. Developments in
agriculture, livestock husbandry and fisheries since past five decades have
attempted to keep pace with the burgeoning population base in Bangladesh.
Especially the agricultural progress witnessed in the world over in the past four
decades has been impressive. The food and fiber productivity improved due to
adoption of innovative technologies, viz. adoption of high-yielding varieties of
crops, irrigation, increased fertilizer and use of plant protection chemicals, mechanization of farm operations and other technology-intensive practices coupled with
public policies favoring maximizing production. Even the committed critics of
green revolution should agree to the fact that the green revolution witnessed in
Bangladesh enabled the country to overcome serious shortage of food and a pre-
carious ship to mouth existence. The technologies were input intensive which were
mostly off farm and synthetic in nature, which led to the deterioration of soil health,
ground and surface water quality and narrowing of the natural resource base. The
revolution was uneven in extending its perceived benefits and excluded a large
chunk of farm community struggling to survive on poor resource base. Continuous
mining of natural resources over the decades no doubts resulted in increase in
production but at the same time lead to degradation of natural resources. As a result,
the productivity levels achieved have not been wholly sustainable ecologically,
environmentally and economically and non-remunerative as well. Reduction of
resource base on one hand and increase in cost of production/cultivation on the
other necessitated the importance of protecting and harnessing natural resources
(Fig. 4.1). The environmental protection movements leaded by the environmental
activism worldwide succeeded in bringing in to focus issues such as soil and water
degradation, narrowing of genetic base, pollution with plant protection chemicals,
nitrification of water bodies and increasing costs of production. Bangladesh also has
very proactive environmental activism highlighting these issues.

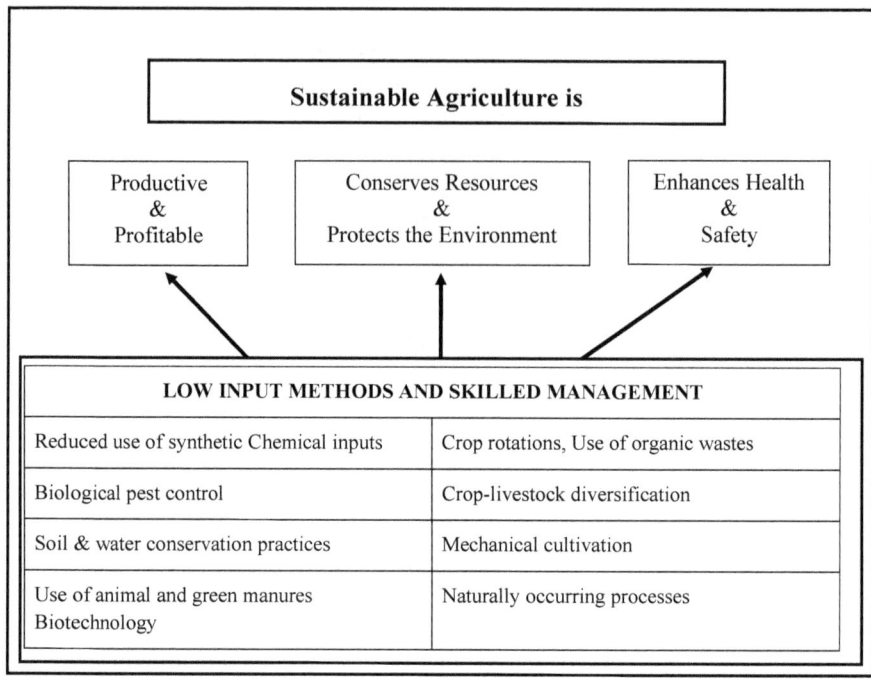

Fig. 4.1 Goals and components of sustainable agriculture

Concept of Sustainable Agriculture

Sustainable agriculture is a concept of farming directed toward maintain or improving the natural resource base while harnessing the same for the production of food, fiber and feed as well as other agro based commodities including livestock, fisheries and aquaculture products. The practices associated with sustainable agriculture involves production of clean foods and other products while sustaining soil health and productivity, water quality, biodiversity and above all economically viable and socially acceptable. The sustainable agriculture practices should have minimum or no negative impact on ecology, environment and quality of human life.

There are different concepts of sustainable agriculture, but none is generally accepted. It embraces several forms of non-conventional agriculture that are often called organic, alternative, ecological or low input. However, from economic and ecological perspectives, two basic criteria must be met if agriculture has to be sustainable in the long term. These are.

Sustainable farming uses some form of integrated pest management for pest control, and this can include the use of chemical pesticides that are not used by organic farmers. Thus, sustainable agriculture does not mean a return to the farming methods of the late 1800's. Rather, it combines traditional techniques that stress conservation with modern technologies, such as improved seed, modern equipment for low-tillage practices, integrated pest management that relies heavily on principles of natural or biological control weed control that depends on crop rotations and manual weeding. Sustainable farms use wind or solar energy instead of purchased energy and use organic animal manure and nitrogen-fixing legumes as green manure to maintain soil fertility, as much as possible, thereby minimizing the need for purchased inputs. The uses of genetically engineered crop varieties are not excluded by sustainable farming. The emphasis is on maintaining the environment, not on rules about what can or cannot be done. Profits from sustainable farms can exceed those of conventional farms.

Innovative farmers have developed many alternative farming methods and systems. These systems consist of a wide variety of integrated practices and methods suited to the specific needs, limitations, resource bases and economic conditions of different category of farmers. To make wider adoption of sustainable agriculture, farmers need to receive information and technical assistance in developing better management skills.

Definitions Sustainable Agriculture

Sustainable agriculture recommends a range of practices, which address many problems that arise due to the problems of modern agriculture such as loss of soil productivity, impacts of agricultural pollution, decreased income due to high production costs, and minimal or uneconomic returns.

Lockets (1988): Defined sustainable agriculture as a time dimension and the capacity a farming system to endure indefinitely.

Gracet (1990): Defined sustainable agriculture as a system of agriculture that is committed to maintain and preserve the natural resource base of soil, water and atmosphere ensuring future generations the capacity to feed them with an adequate supply of safe and wholesome food.

Crosson (1992): A sustainable agriculture system is one that can indefinitely meet demands for food and fiber at socially acceptable economic and environment cost.

However, commonly accepted definition of sustainable agriculture as production and distribution system that:

- Achieves the integration of natural biological cycles and controls
- Protects and renews solid fertility and natural resource base
- Reduces the use of non-renewable resources and purchased (external and off-farm inputs) production inputs
- Optimizes the management and use of on-farm inputs
- Provides an adequate and dependable income
- Promotes opportunities in family farming and farm communities

- Minimizes adverse impacts on health, safety with life, water quality and the environment
- Provides on-farm employment to the rural small and marginal farmers.

A comprehensive definition of sustainable rural development including farming systems offered by FAO in 1988 as: "Sustainable rural development is the management and conservation of the natural resource base and the orientation of technological and institutional change in such a manner so as to assure the attainment and continued satisfaction of human needs for the present and future generations. Such sustainable development, in the agriculture, forestry and fishery sectors, conserves land, water, plant and animal genetic resources, is environmentally non-degrading, technically appropriate, economically viable and socially acceptable."

Goals of Sustainable Agriculture

Sustainable agriculture is any system of food or fiber production that systematically pursues.

- A incorporation of natural processes such as nutrient cycling, biological nitrogen fixation and pest predator relationships into agricultural production processes
- A reduction in the use of off-farm, external and non-renewable inputs (synthetic) with the greatest potential to harm the environment or the health of farmers and consumers and a more targeted use of remaining inputs used with a view to minimize variable costs
- Access to productive resources and opportunities, and progress toward more socially just forms of agriculture
- Use of biological and genetically potential of plant and animal species
- A judicious match between cropping pattern, environmental constraints of climate and landscape to ensure long-term sustainability of current productivity levels
- A greater and productive use of local knowledge and practices, including innovation in approaches not yet fully understood by scientists but widely understood and adopted by farmers
- A full participation of farmers and rural people in all processes of problem analysis and technology development, adoption and extension
- Profitable and efficient production with an emphasis on integrated farm management and conservation of soil, water, energy and biological resources.

The ultimate goal of sustainable agriculture is to maximize the benefits from the existing agricultural assets and minimize the threats to the environment from the current practices of technology-intensive agriculture directed only toward production maximization and profits.

Differences Between the Contemporary and Sustainable Types of Agriculture

Contemporary Agriculture

- System simplification/monoculture
- Specialized agriculture (economically vulnerable)
- Soil is considered just as a medium
- Feed the plant directly but not the soil
- Linear flow of nutrients/energy
- Reductionist.

Sustainable Agriculture

- System of production diverse not restricted by monoculture of crops
- Recycling of nutrient pool for crop
- Feed and nurture the soil and not the crop
- Holistic approach to farm productivity.

Positive Aspects and Factors Which Enhance Modern or Sustainable Agriculture

Modern Agriculture

- High yield/high and fast returns/profit oriented
- Increased mechanization
- Scope for intensive cropping
- Better compatibility through genetic homogeneity in inter/mixed cropping
- New varieties of plant breed (pest/disease tolerant)
- Maximum utilization of land and water
- Meeting the need of sufficient and fast food production
- Caters to habituation of food through monoculture
- Immediate and direct supply of nutrients to the plant through chemical fertilizers (NPK). Quick responses to inputs, fertilizers, chemicals
- Better pest, disease, weed control

- Package of practices for different locations, situations, crops, agro climatic regions
- Targeted yields evolved and recommended.

Sustainable Agriculture

- Affordability by any farmer
- No sophisticated/imported and special technology is necessary
- Environmental conservation and protection
- Healthy atmosphere/healthy food
- Prevent/avoid ecological degradation
- Increasing stability and status of soil fertility
- Security more through higher levels of disease and pest resistance
- Recycling of nutrients
- Substance of soil fertility through organic recycling
- Diversity
- Inter dependency
- Efficient use of natural resources
- Self-sustaining.

Negative Aspects, Problems, Barriers and Issues

Modern (Technology Intensive) Agriculture

- Short-term benefit, operates law of diminishing returns
- Depletion of nutritional base of the soil, water and atmosphere quality
- Environmental pollution due to use of chemicals (water, soil, environmental)
- Health hazards due to entry of pesticides, toxins, antibiotics, heavy metals in to food chain
- High cost of production
- Increasing dependency on external inputs
- Less diversification manifested through disappearance of genetic races because of monoculture leading to risks such as loss of biodiversity, pest and disease resistance and resurgence due to production of "unclean" food from the overall public health point of view
- Poor quality of produce
- Economic disparity in the society widens. Rich becoming richer and poor becoming more poorer

- Operates against principles of nature and ecology
- Natural parasites, predators and beneficial insects are adversely affected and totally disappear over a period of time.

Sustainable Agriculture

- Takes longer time to realize the benefits of regenerative farming
- The change is gradual
- Relatively difficult to motivate farmers for change initially but once convinced adoption easy
- Comparatively labor-intensive—needs proper planning for allocation/use of available resources
- Initial yield is low.

Lack of infrastructures and systems for documentation of indigenous knowledge and techniques, and inappropriate extension services for propagation. Lack of recognition, acknowledgement, motivation and incentives from the state. Financial assistance and back up, particularly to a small farmer is low, in comparison to what is available for adopting modern agriculture (subsidies inputs).

Elements of Sustainable Agriculture

Sustainable agriculture consists of elements, which are common in many regions. But the methods to improve their sustainability may vary from one agro ecological region to another. However, there are some common sets of practices among farmers trying to take a more sustainable approach by use of on-farm or local resources. However, each of them contributes to a greater extent to realize long-term farm profitability, environmental stewardship and quality of life.

(A) **Soil conservation**: Soil conservation methods including contour cultivating, contour bunding, graded bunding, vegetative barriers, strip cropping, cover cropping, reduced tillage, etc. help prevent loss of soil due to wind and water erosion.

(B) **Crop diversity**: Increased crop or biodiversity on farm can help reduce risks from extremes in weather, marketing conditions and pest disease incidences. The increased diversity of crop and other plants such as trees, shrubs and pastures also can contribute to soil conservation habitat protection and increased populations of beneficial insects.

(C) **Nutrient management**: Integrated management of essential nutrients can improve and sustain soil fertility and protect environment. Increased use of on-farm low-cost inputs such as organic manures, composts, green manures and crop residues not only reduces cost of production but also rejuvenates soil health.

(D) Integrated pest management (IPM): It is a sustainable approach to manage pests by aptly integrating the available plant protection methods like cultural, physical, mechanical biological and chemical methods, which optimizes the production costs besides maintaining environmental balance.

(E) Water quality and water conservation: Practices like zero tillage, deep plowing, and mulching and microirrigation techniques and mulching can help to optimize the water consumption or requirement besides conserving and augmenting the soil moisture on long-term basis. It also helpful in protecting the quality of drinking water and surface water.

(F) Agro forestry: A combination of silvipastoral, agri-silvipastoral, agri-horticulture, horti-silvipastoral, alley cropping, ley farming, etc. that can help conserve soil and water and profitability. Also leads to supply of fuel wood, horticultural products and achieve balanced nutrition to rural people.

(G) Marketing: Improved marketing facilities can ensure remunerative and sustainable returns to farmers. Direct marketing of produce can exclude intermediaries and ensures higher returns and malpractices.

Low External Input Sustainable Agriculture (LEISA)

Low external input supply agriculture (LEISA) is a component of sustainable agriculture. It can be defined as production activities which optimize the use of locally available resources by maximizing the complementary and synergetic effects of different components of farming system.

LEISA is based on a preventive approach wherein the problem is tackled at its roots as opposed to the more symptom curing nature of modern/chemical agriculture. LEISA is more labor intensive and often based on local knowledge and production systems.

Criteria for LEISA

Ecological Criteria

- Balanced use of nutrients
- Efficient use of water, energy and genetic resources
- Minimal/need-based external inputs
- Minimal negative environmental impact.

Economic Criteria

- Sustained farmer livelihood system
- Competitiveness
- Efficient use of production factors
- Low relative value of external inputs.

Social Criteria

- Widely acceptable and equitable adoption potential especially among small farmers
- Reduced dependency on external institutions
- Respecting and building ITK, beliefs and value system
- Contribution to employment generation.

Factors Influencing Ecological Balance in Sustainable Agriculture

Major factors which influence the resource base and ecological balance in sustainable agriculture are.

Soil-Related Factors

- Accelerated soil erosion and degradation
- Deforestation
- Siltation of reservoirs.

Irrigation-Related Factors

- Rise in ground water table and water logging
- Soil salinization and alkalization

- Overexploitation of ground water and reduction in ground water resources, i.e., depletion of water table.

Agro Chemical Pollution

- Fertilizer pollution
- Pesticide pollution.

Environmental Pollution

- Greenhouse gases
- Impact on ozone layer
- Methane emissions from soil and livestock husbandry.

Characteristics of Bangladesh Traditional Agriculture

Traditional agriculture or subsistence-based agriculture is primarily based on getting the most yield possible from a small plot of land. Since most of the land is inherited and is divided among successors, subsequent generations receive smaller and smaller plots of land. This means that to maintain a steady source of yield, more intensive techniques need to be employed. Thus, the use of both organic and chemical fertilizers is employed, irrigation is used, additional labor force is employed, etc. This is to squeeze as much yield as possible from a limited amount of land area.

For this reason, cropping intensity in Bangladesh in extremely high at around 179%. This refers to how much yield can be obtained from a certain amount of land. Moreover, around 56% of agricultural land is irrigated. While there are many river networks to help water the farms, these networks are not spread evenly across the country. Thus, the drier regions benefits from having irrigated water supply.

The intensive nature of agriculture is evident from the land-to-man ratio of only 0.06 ha. Most of this land is inherited, and this results in a very fragmented land supply. Fragmentation results in wasted land near the edges due to the presence of hedges, fences or barriers and thus the usable land area is even smaller.

One interesting point of note is the social structure around rural communities. Historically, they had a large family with many children. A large family meant that more labor force to help out with agricultural activities which meant that a higher output could be extracted from the same plot of land. However, when it came to inheritance, the same plot of land had to be divided among many individuals which reduced future output potential as the land became fragmented.

This high dependence on manual labor is due to the fact that most of these farmers are not rich enough to afford modern machineries and mechanized farms. Instead, these farms are run through the use of simple equipment and are mostly animal driven. The manure from the domesticated animals are used as fertilizers and if the soil quality is poor then chemical fertilizers are used. However, reinvestment into the farms tend to be low as one seasons output can be used as the next seasons input. Reinvestment is also kept low in order to minimize expenditure as the output from these farms tend to produce few surplus. Whatever surplus are made though, are sold at the local market. Since these are unprocessed outputs, they do not fetch high prices and thus are sold cheaply. Furthermore, the lack of differentiation between sellers mean that competition can be fierce which keeps the costs low. All of this means that farmers do not earn a lot from subsistence farming.

Agriculture and the Environment in Bangladesh

Depending on the methods employed, agricultural activities may cause severe soil deterioration. Some techniques such as mixed cropping and fallowing does leave room for soil quality to improve over time but practices such as monoculture rapidly depletes the soil of its nutrients. As nutrients in the soil are in fixed supply, having a large number of similar species will focus too deeply on select resources, depleting it quickly and diminishing the health of the system in the process. Evidently, areas where monoculture are practiced, a large inflow of artificially introduced nutrient supply such as fertilizers are required at regular intervals to replenish the lost nutrients.

Agricultural practices also release high amounts of greenhouse gasses and is responsible for around 13% of total global emission. This makes it the second largest emitter after the energy sector. Majority of this emission are in the form of methane and nitrous oxide. The sources of these are cattle belching, burning of crop residue, manure management, etc.

There are further indirect effects of agricultural on the environment too. These effects are mostly from how agriculture is set up. Oftentimes, patches of forest areas are cleared out to free up land for cultivation. This means that the loss of photosynthesizing and oxygen releasing trees must be taken into account as losses in carbon sink. Burning of forest patches also releases carbon dioxide. Excess fertilizers are also potential sources of emission.

In addition to all of these, there are several other environmental problems associated with agriculture. These are:

Deforestation: This results from the lack of land area needed to provide adequate food supply to the local inhabitants. Everybody relies on crops such as rice and wheat. To cultivate these if adequate land area is not readily available, forest patches are usually cleared to provide space for agricultural practices.

Desertification: This compounds with deforestation. Some ecosystems are highly sensitive to change. When areas are cleared, these land areas lose their balance and their nutrient cycle gets disrupted. With agriculture being done in such areas and with the use of fertilizers and irrigation, the soil health degrades rapidly to a point where it cannot sustain further growth and is unable to support floral diversity. At this point, the land slowly adopts desert like conditions. Furthermore, without an intricate root network to bind the soil together, the soil becomes susceptible to wind erosion.

Soil erosion: This is when the soil becomes loose and is easily removed by forces such as wind or rain. The topsoil holds most of the nutrient, but is also easily disturbed by excessive rainfall or strong wind action. Loss of topsoil results in poorer soil quality which causes a feedback loop that spirals into poorer and poorer soil health.

Overgrazing: This is especially highlighted in the "tragedy of the commons." Having a large number of organisms feeding on a small plot of land removes too much of the grass cover. This loosens up the soil considerably and halts future grass cover growth. The land becomes barren and unproductive as a result.

Water related pollution: Large quantity of water resources are needed to maintain peak crop yield. Depending on the geographic context of the farmlands, river water may not be within reach. Thus, farmers usually resort to using irrigated water. However, this cause salinization which deteriorates soil health.

Also, use of excessive fertilizers leaves an abundance of phosphorus, nitrates, potassium, etc. This, coupled with monoculture, means that not all of the nutrients are used up in the same ratio. Thus, during periods of heavy rainfall, the excess nutrients run off and flow to nearby river networks. The inflow of excessive nutrients causes algal bloom or eutrophication. Eutrophicated river surfaces deprive the bottom of the river from sunlight, which can be deadly for aquatic floral and faunal species.

Effects of Climate Change in Bangladesh Agriculture

In Bangladesh, food security has been one of the major national priorities for last few decades but the target has always been interrupted by the climate change and for resource constraints. Present section of this chapter will highlight the major effects of climate change in the food production and the national resources constraints to address the food security.

However, major constraints in terms of food security in Bangladesh attributed to cultivable land scarcity, irrigation water scarcity in summer, lack of technological knowledge, lack of climate adaptive crop variety, lack of institutions and professionals as well as social and cultural constraints are prominent.

1. Impact of temperature on crop production
2. Impact of rainfall on crop production
3. Impact of sea level rise on crop production
4. Impact of flood on crop production
5. Impact of drought on crop production
6. Land scarcity
7. Irrigation water scarcity
8. Lack of technological knowledge
9. Inadequate institutions and professionals.

In order to mitigate the adverse impacts of climate change on food sector, we need to analyze the possible options that could assist in increasing food security. Therefore, adaptation in the agriculture sector must be well integrated with both the broad national development goals and livelihood priorities at the local level. Rural agrarian people have long been adapted to a variety of climate risks with their traditional knowledge. These coping strategies are varied depending on regions and prevailing socio-economic conditions. As the climate change is a reality now, more and different adaptation intervention is required to ensure food security within a given time.

Formal and informal sources of support can play critical role in minimizing climate risks on food security. The supports may be investments in agriculture and water resources, or may be on infrastructures (e.g., embankments in floodplain and coastal areas to protect against floods and storm surges) or irrigation.

Groundwater irrigation plays an important role in crop agriculture in the drought-prone areas. Irrigation provides a mean to adapt soil moisture condition with diversifying crop agriculture, promoting high-yielding variety crops and increased cropping intensity. Flood-prone areas of the Southern Bangladesh coastal embankment provide protection to crop agriculture and livelihood assets playing a great role in food security. In recent years, government of Bangladesh has invested over USD 10 billion (at constant 2007 prices) for flood management in embankments, coastal polder and cyclone shelters (BCAS, personal communication). With this protection, substantial increases in production have been made possible.

Status of Sustainable Agricultural Systems in Bangladesh

The strides made by Bangladeshi Agriculture in the past four decades have been impressive. Food grain production has increased meeting the food needs of the burgeoning population. However, the main spurt in production has been in irrigated agriculture sector mainly in wheat and rice crops. These are the crops which received the greatest stimulus of varietal improvement, shared major proportion of irrigation and fertilizers and benefited most from governmental price support and procurement policies. In contrast pulses, oilseeds and coarse cereals which are cultivated mainly in the rain-fed sector remained rather static and deprived of these benefits and are pushed to marginal areas.

Seventy percent of the cultivated area in the country lies in the region of medium to low rainfall (1150 mm and below), and in most cases, the rainfall is inadequate and uncertain with respect to crop water requirements. Higher level of crop production and its stability can be obtained only through irrigation. Till 1951, when the first five-year plan was launched, Bangladesh had an irrigation potential of 22.6 mha. The eight plan target provides for the creation of additional 15 mha potential of which 13.5 mha would be the target for utilization in the terminal year. The expansion of irrigation resources brought about not only enhancement in production but also stability in production.

Nutrient management is the key to higher crop productions. Most of the growth in the food production during the green revolution period is attributed to the higher fertilizer use. The annual fertilizer consumption is expected to rise to about 20 million tons in Bangladesh by the end of this century. This rise in fertilizer use is necessary because we foresee (i) N deficiency will continue to be universal in Bangladeshi soil, (ii) deficiency of P will be next to the order of the extent, (iii) K will become limiting in high-productive regions, (iv) in at least half of the Bangladeshi soils, crops would benefit from Zn treatment and (v) S deficiency will obstruct the optimum productivity in vast majority of Bangladeshi soils. We do not rule out the emergence of additional nutrient disorders, which may stand in the way of sustaining the envisaged growth rate in agricultural productivity. The situation, therefore, calls for an integrated nutrient management and supply system which (i) integrates the use of on-farm generated organic manures with inorganic fertilizers and (ii) exploits the natural nutrients supply through bioengineering and mining of soil resources.

With the advent of the green revolution beginning 1965, the new crop varieties and cropping sequences for intensive agriculture brought to the fore front problems of pests and diseases which caused losses to various crops and their produce. Pesticide consumption has shown a steady increase in Bangladesh. From the mid-fifties, application has increased both in quantity and in coverage. Plant protection coverage increased from 2.4 million hectares in 1956 to 80 million hectares in 1984 and the quantity of pesticides rose from 200 tons in 1955 to 72,000 tons (tech.) in 1987. Their influence in upsetting the ecological balance and polluting the environment is being felt increasingly. Safe use of pesticides should therefore, be the target.

Good-quality seed is one of the basic requirements for augmenting crop production and productivity. The extent of area under high-yielding varieties (HYV) is one of the important indicators of the spread of modern technology for increasing agricultural production. The scope of exploiting the genetic potential of plants for higher yields under different input intensities and under different situations is tremendous. Development of appropriate crop varieties is going to be the main component of sustainable agriculture.

The research strategy for sustainable agriculture should be an exercise in the development of such farming systems which meet the production objectives through most efficient utilization of inputs without impairing the quality of environment with which the system interacts. Use of land according to its capability, integrated use of purchased inputs and farm residues of production, harnessing maximum productive efficiency from the inputs and from the interacting environments and considering these objectives in long-term perspectives are some of the important aspects of sustainable agriculture strategy.

The need is to integrate the components and evaluate the synthesized systems against the existing system of production. Long-term monitoring of the improved systems with regard to the parameters of sustainability will be required. On-station research could be initiated but the on-farm testing of the developed technology will be required ultimately. The following may be the major course of research work in future on sustainable agriculture in Bangladesh.

- Synthesis of the sub-systems of sustainable agriculture through on-station research
- Evolving a system through synthesis of good production practices for a sustainable farming system based on the location specific needs through on-farm research
- Development of research methods for evaluating the farming systems with regard to its biological, environmental and social efficiency
- Development of methodology for monitoring the improved systems over long period of time.

Issues Related to Gender in Sustainable Agriculture

- Scope of women participation in sustainable agriculture
- The role of women in agriculture is not expressed properly
- Development has neglected women
- Contribution of women to agriculture is not recognized
- Access to agricultural assets. Rights on ownership land/property
- Social attitudes toward women participation
- Low wages for women in agriculture
- Overburdened with work
- No access to appropriate training
- Poverty and its overbearing on women (in agriculture)
- The need for equal participation of men and women for sustainability in agriculture
- Gender bias in agricultural technology generation.

Challenges in Bangladesh Agriculture

The importance on agricultural output cannot be overstated as it forms the carbo-hydrate source for almost the entire population of Bangladesh, and for people all over the world. However, its environmental impact is also non-negligible. There are many issues associated with agriculture ranging from environmental to societal. A sustainable system for agriculture must work to address these issues. Environmental issues are becoming more evident by the day, with greenhouse gas emission rising and deterioration to soil and water quality getting worse. There are societal issues as well, such as modernization of agricultural practices leading to rising inequality. Richer farmers can afford modern machineries, genetically modified crop seeds and chemical fertilizers. This results in efficient farming, higher yield from a unit plot of land or better utilization of larger farmlands. This not only produces good quality and large quantity of output, but it also means that the produces can be sold at cheaper rates at the local market, or can be processed and exported for higher rates. Farmers who cannot afford such technology are at a disadvantage. With an ever worsening climate, it is difficult to sustain the same amount of yield as they could previously. This issue is compounded with land fragmentation and unsustainable farm practices. With poor quality output and low quantity available to be sold, it becomes difficult to compete with richer farmers, landing them at a competitive disadvantage. This only serves to widen the inequality present.

The aim of sustainable agriculture is to promote production while minimizing environmental impact. However, there are many challenges that needs overcoming in order to achieve this goal. Agricultural area is on the decline at the rate of 1% per annum. However, population growth remains at 1.48% per year. To feed the ever growing population, more farm area is needed. Food shortages might become more apparent as climatic conditions are worsening which means poorer quality output and food shortages are likely. Rapid urban growth also puts a strain on agricultural land as rural areas which are suitable for farming are converted to pavements and concrete surfaces.

There are also many technological challenges that lie ahead, mostly in terms of new innovation and disseminating existing expertise and knowledge. More research is needed in making agricultural output more efficient and less impactful on the environment. Lastly, focus should be directed at restoration of soil quality and fertility.

Conclusion

The fast development of technology for increasing production without giving due importance to the agro ecosystem balance resulted in disturbed biological rela-tionships. The imbalances thus created lead to fast degradation of natural resource base. Thus, the present productivity levels have become unstable and uneconomic. This necessitated for maintenance of natural resources so as to meet future demand. Thus, the concept of sustainable agriculture emerged.

The sustainable agriculture concept lies in the successful management of resources for agriculture to satisfy changing human needs while maintaining or enhancing the natural resource base and avoiding ecological and environmental degradations.

Sustainable agriculture aims at production of safe and clean food without harming the quality of natural landscapes and with minimal impact on environment. However, it should operate within socially acceptable system and economic viability.

Number of elements right from soil water, biota to agro ecosystems if not judiciously used and maintained may lead to disasters in years to come. Then, they become major factors which can hamper the sustainable agriculture systems.

The backlash of modern agriculture in terms of degradation of natural resource and economical base of farming system compelled to look at low external input supply agricultural systems where in local resources, knowledge, social values are safeguarded with economic viability.

Bibliography

Adger WN, Agrawala S, Mirza MMQ, Conde C, O'Brien K, Pulhin J, Pulwarty R, Smit B, Takahashi K (2007) Assessment of adaptation practices, options, constraints and capacity. Climate change 2007: impacts, adaptation and vulnerability. contribution of working group II to the fourth assessment report of the intergovernmental panel on climate change, pp 717–743. Cambridge University Press, Cambridge, UK

Ashraf MY, Sarwar G, Ashraf M, Afaf R, Sattar A (2002) Salinity induced changes in α-anylase activity during germination and early cotton seedling growth. Biologia Plantarum 589–91

ASR (2006) Agriculture sector review (crop subsector). Actionable policy brief and resource Implications. Ministry of Agriculture, Govt. Republic of Bangladesh, Dhaka, pp 14–51

Bangladesh Ministry of Agriculture (MoA) (2012) Available at http://www.moa.gov.bd/statistics/bag.htm. Accessed on 7 Aug 2012

Basak JK (2009) Climate change impacts on rice production in Bangladesh: results from a model. Published by Unnayan Onneshan—The Innovators, Dhaka, Bangladesh

BBS (2008) Statistical year book of Bangladesh. Bangladesh Bureau of Statistics Division, Ministry of Planning, Govt. People's Republic of Bangladesh, Dhaka, pp 121–134

BER (2009) Bangladesh economic review. Department of Finance, Ministry of Finance, Govt. People's Republic of Bangladesh, pp 83–85

Bharadwaj KK, Gaud AC (1985) Recycling of organic wastes, p 104. ICAR, New Delhi

BRRI (2010) Rice statistics in Bangladesh. Bangladesh Rice Knowledge Bank, Bangladesh Rice Research Institute, Gazipur. http://www.knowledgebank-brri.org/riceinban.php

Chowdhury MR (2009) Population challenge facing Bangladesh. Long Island University, CW Post Campus, New York. http://www.fao.org/fileadmin/user_upload/fisheries/docs/Rice_Fish_Farming_Bangladesh.pdf

Danama AK. Organic farming for sustainable agriculture

DFID (2001) Sustainable agriculture evaluation. Bangladesh Country Report. 7, 85

DFID (2002) Sustainable agriculture. Resource management key sheet 10. DFID, London UK

DoF (2010) Fisheries statistical yearbook of Bangladesh 2008–2009. Fisheries Resources Survey
    System, Department of Fisheries, Ministry of Fisheries and Livestock, Dhaka, p 1
http://www.attra.org
http://www.hawiaa.org/iaeq.htm
http://www.infohabitat.org/treaties
http://www.qgroecology.org/principle/ecosustdef.htm
http://www.sare.org/pulications/exploring.htm
http://www.sustainabletable.org
NAEP (1996) New agricultural extension policy. Ministry of Agriculture, Govt. People's Republic
    of Bangladesh, pp 1–10
NAP (1999) National agricultural policy. Ministry of Agriculture, Govt. People's Republic of
    Bangladesh, pp 1–23
NSP (1993) National seed policy. Ministry of Agriculture, Govt. People's Republic of Bangladesh,
    pp 1–10
Pretty (1995) Regenerating agriculture. Earthscan, London, UK
Pretty (2002) Agri-culture: reconnecting people, land and nature. Earthscan, London, UK
Rao VP, Veer Raghavaiah R. Manual on farming systems and sustainable agriculture
Sharma AK. Handbook of Organic farming
Stark CR (1995) Adopting multidisciplinary approaches to sustainable agriculture research:
    potentials and pit falls. Am J Altern Agri 10(4):180–183
USDA Farm Bill (1990) United States Congress, 1990. Food, Agriculture, Conservation, and
    Trade Act of 1990, Public Law 101–624. Title XVI, Subtitle A, Section 1603. US Government,
    Washington, DC

Chapter 5
Climate Change and Its Impact on Health in Bangladesh

Mohammad Delwer Hossain Hawlader

Abstract Bangladesh is one of the most vulnerable countries of the world to extreme weather, mainly due to its huge coastal area, highest density of population, high rate of poverty and lack of natural resources. Day by day, the quality and the quantity of our natural environment have been deteriorating, and the main reasons for these changes are anthropogenic. Climate changes are expecting to contribute to some air quality problems. Respiratory diseases may be exacerbated by warming-induced increased frequency of events and allergen in air. It has been said that ground-level ozone can damage lung tissue, and it is harmful for those who have asthma and other chronic lung diseases. The preparation of Bangladesh to face the challenge of global warming is not enough and cannot be overlooked. In order to tackle the health and socio-economic effects, relevant stakeholders including policy-makers, program designers, program implementers, civil servants and civil society members need to have a better understanding of both climate change and its possible impacts. Appropriate and relevant policy need to be formulated and follow-up for proper implementation. In recent past, Bangladesh has many success stories especially in health sector and human development index to share with the rest of the world. Through proper understanding of climate change issue and its impending effects on health, timely initiation of preparedness activities with thorough participation of community might helps to mitigate the problems.

Keywords Climate change · Temperature variation · Chronic diseases · Vulnerability · Bangladesh

M. D. H. Hawlader (✉)
Department of Public Health, North South University, Bashundhara R/A, Dhaka 1229, Bangladesh
e-mail: mohammad.hawlader@northsouth.edu

© Springer Nature Switzerland AG 2021
Md. Jakariya and Md. N. Islam (eds.), *Climate Change in Bangladesh*,
Springer Climate, https://doi.org/10.1007/978-3-030-75825-7_5

85

Introduction

Climate change is becoming a major concern globally, especially for its impact on the global burden of disease, disability and premature deaths. Climate change is affecting human health either directly or indirectly or even both (WHO climate and health country profile 2015). All of us are directly exposed to the different types of climate change patterns, such as temperature, precipitation, sea level rise, frequent extreme events and also indirectly exposed through changes in the quality of water, air and food, as well as changes in ecosystems, agriculture, industry, human settlements and finally the economy (WHO climate and health country profile 2015; Kjellstrom et al. 2018). These direct and indirect exposures are causing morbidity, mortality, disability and human suffering. These above-mentioned health problems gradually increasing the vulnerability and reducing the capacity of individuals and groups to adopt these changes (Grecequet et al. 2017). Bangladesh is one of the most vulnerable countries of the world to extreme weather, mainly due to its huge coastal area, highest density of population, high rate of poverty and lack of natural resources. Day by day, the quality and the quantity of our natural environment have been deteriorating, and the main reasons for these changes are anthropogenic. In the last few decades, climate change has huge impacts on human lives and livelihoods by increasing frequency and intensity of various events of disasters (Hasib and Chathoth 2016), and nowadays, it is becoming a major concern all over the world (Fig. 5.1).

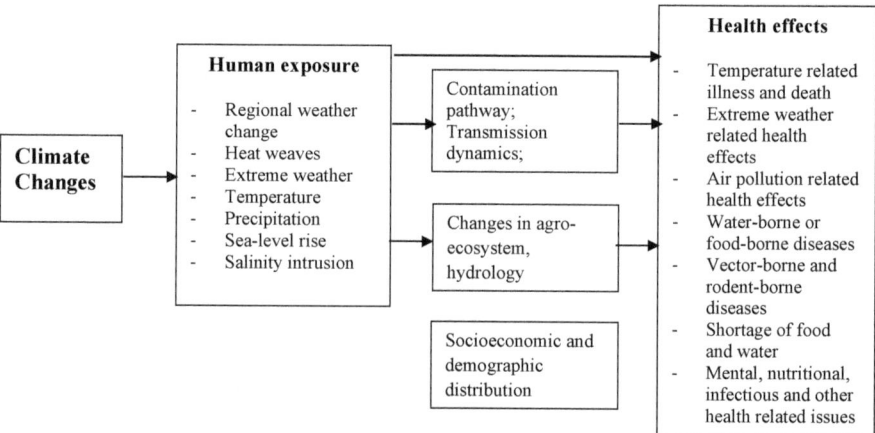

Fig. 5.1 Relationship between climate change and human health. *Source* WHO 2003

Climate Change in Bangladesh

In the last decade, Bangladesh was hit by two consecutive devastating cyclones within two years gap, which are Sidr and Aila. Among many cyclones in Bangladesh history since 1965, most recent one was Aila in 2009 which caused 400 deaths, but Sidr in 2007 hit Bangladesh on mid-November 2007 and caused almost 3500 deaths and over 55,000 injuries (Rahman 2008). But history of some other grave cyclones has been found in the record of Bangladesh Meteorological Society. Those cyclones were occurred in the last 5/6 decades where many life along with natural resources has been lost. Most dangerous one found in the history was in 1970, where almost 300,000 people died with many more casualty recorded. The second highest death recorded from 1991 cyclone, where 138,882 death occurred with many injuries. Now, it is predicted that more than 20 million people will be living in cyclone risk areas by 2050 compared to 8.3 million at present (Table 5.1).

Above-mentioned cyclones caused many deaths and disabilities. Still now, country is facing tremendous difficulties to rehabilitate those cyclone-affected people. Not only cyclone, recent rise of temperature is inviting many communicable diseases which is creating double burden of disease. It is estimating that the annual mean temperature is projected to be raised by around 4.8 to 5.0 °C within 2100. This temperature rise is causing sea level rise every year, which causing erosion and submerging the coastal area and ultimately increasing the number of homeless population and will cause social disruption (Gilman et al. 2007). Those homeless populations struggle to maintain their healthy life and ultimately increase disease burden. Country needs to take proper intervention right now; otherwise, an annual average of more than 7 million people are anticipated to be affected by flooding due to sea level rise between 2070 and 2100. Climate change can play an important role in the change of disease transmission and growth of vectors for some infectious

Table 5.1 Major cyclones over Bangladesh at a glance

Date	Year	Maximum wind speed (km/hr)	Storm surge height (metre)	Death toll
11 May	1965	161	3.7–7.6	12,297
15 December	1965	217	2.4–3.6	873
1 October	1966	139	6.0–6.7	850
12 November	1970	224	6.0–10.0	300,000
25 May	1985	154	3.0–4.6	11,069
29 April	1991	225	6.0–7.6	138,882
19 May	1997	232	3.1–4.6	155
15 November	2007	223	6.1–9.1	3,363
25 May	2009	170	5.2–10.0	400

Bangladesh Meteorological Society 2009

diseases (Waits et al. 2018) and some allergenic pollen species (Barnes 2018). Temperature raise has an important role in the seasonal pattern or distribution of incidence of malaria, dengue, tick-borne diseases, cholera, and other diarrhoeal diseases (Phung et al. 2018). It is assuming that by 2070, over 147 million people are projected to be at risk of malaria. Climate also may cause malnutrition due to lack of production of food because of drought. Even though Bangladesh contributing the lowest amount of emission which driving global warming, it is one of the worst victims of global warming effects (Kabir et al. 2014).

Health Aspect in Relation to Climate Change

Climate change has hundreds of effects on human health. The most common health outcomes are diarrhoeal disease, respiratory infections, allergy, fatal accidental injuries due to floods and landslides, malnutrition and vector-borne disease like dengue, malaria, kala-azar and chikungunya. Increase in extreme events such as cyclone, flood and earthquake is most likely to cause huge number of injuries and death (Kabir et al. 2016). Several vector-borne diseases are worthy to mention. Nowadays, the most common one is dengue, which is becoming the world's most important vector-borne viral disease, and it was not prevalent in Bangladesh until 2000. Since then, it becomes epidemic, and now, we can say it became endemic in urban areas because of its constant presence with high number. The incidence of dengue occurs every year in Bangladesh, especially in major city, more predominant in Dhaka city, and it is creating constant threat to the population and is a recurring problem for the health authorities and city corporation authority. Moreover, all environmental conditions that can trigger an outbreak of infectious disease are some extent in recent time (Kabir et al. 2016; Elahi and Khan 2015). Another epidemic like Chikungunya is creating a huge concern in the last few years, especially throughout Dhaka city and some other large cities of Bangladesh. For both dengue and chikungunya fever, the transmission usually starts during June–July every year with some seasonal fluctuation. Its seasonal peak is observed during mid-July when the temperature and humidity are in its extreme (Choudhury Zamil et al. 2008). Most of the cases of dengue and chikungunya are observed in Dhaka, Chittagong and Khulna, with the highest incidence rate in Dhaka city. Since mid-nineties, locally known kala-azar (leishmaniasis) is another vector-borne disease that has been re-emerged (Bern and Chowdhury 2006). It is estimated that in more than 27 districts of Bangladesh, around 20 million people are at risk of kala-azar. A total of around seventy-three thousand kala-azar cases were reported to the Directorate General of Health Services (DGHS), Government of Bangladesh, from 1994 to 2004. Malaria is another vector-borne disease, and even after lot of efforts from government and different national and international organizations, it cannot be eliminated due to the pattern of temperature and humidity. Even though it is not equally distributed in all districts, still it is a major public health problem in Bangladesh. Thirteen bordering districts out of 64 districts are at risk of malaria.

Recently, WHO estimated more than 14 million populations are at risk of malaria in Bangladesh (World malaria report (2010)). Mostly, districts of Chittagong Hill Tracts have the highest prevalence rate than the other endemic districts (Ahmed et al. 2008). During flood, there is huge possibility of increased communicable diseases by drinking and contacting with contaminated floodwaters. During summer and after flood, outbreak of diarrhoea occurs almost every year in Bangladesh. Watery diarrhoea among the displaced population due to floods is the most common cause of death for children under 5 years of age (Noé et al. 2018) Vulnerability of diarrhoeal diseases is high among the population of floodplains and river deltas areas. World Bank reported that the diarrhoeal disease contributes almost 10% of the total burden of disease in Bangladesh mainly due to inadequate access to safe water, lack of sanitation and poor hygiene (Mahmud and Mbuya 2015).

As a result of climate change, due to change in rainfall pattern, drought is increased in some areas of Bangladesh. The major effects of drought on health include malnutrition, infectious diseases and respiratory diseases. Drought can cause diminished dietary diversity, reduces overall food consumption and may therefore lead to under nutrition (Menne and Bertollini 2000). Until now, there is a scarcity of formal study focused on increase of heatwaves in Bangladesh, but the increasing trend of temperature has been found in Bangladesh by several recent studies. Increased temperature may result in many complicacy for human health. Study reported that the health impacts associated with heatwaves are heatstroke, dehydration and aggravation of cardiovascular diseases (Rahman 2008). In recent past, it was also observed that prevalence of diarrhoeal diseases increased during extreme temperatures and heatwaves, particularly among the young children (Borroto 1998).

Climate changes are expecting to contribute to some air quality problems. Respiratory diseases may be exacerbated by warming-induced increased frequency of events and allergen in air. It has been said that ground-level ozone can damage lung tissue, and it is harmful for those who have asthma and other chronic lung diseases (Louis and Hess 2008). For other types of pollutants, the effects of climate change are not studied well, and results are varying by region to region. Nowadays, another concern pollutant is "particulate matter," also known as particle pollution or PM, and is a complex mixture of extremely small particles and liquid droplets. When we breathed in those PM, particles can reach the deepest regions of our lungs. Climate change may indirectly affect the concentration of PM pollution in the air by affecting natural or "biogenic" sources of PM such as wildfires and dust from dry soils (IPCC 2007).

Socio-economic Aspect

Recent changes in climate might be affecting natural resources, such as air, water, soil, forests and grasslands. Changes in natural resources will have some detrimental effects on both society and economy (Sellers and Ebi 2017). For example, sea level rise may increase the rate of soil erosion and leaching, which detrimentally affect agricultural resources. These effects on agricultural resources may affect the social and economic circumstances of farmers and other socio-economic sectors depending upon their agricultural production. Therefore, the socio-economic effects of climate change arise from a complex interaction between climate and society and how these affect both natural and managed environments. Climatic variations may provide few opportunities though, but it imposed lot of costs, depending on how society can adapt to these changes. Such as, a bountiful floodplain rice-growing system, is often disrupted by floods, droughts and cyclones. In the future, the extent to which Bangladesh will be affected will depend on the future technological, demographic and socio-economic trends and how they influence Bangladesh's ability to adapt in order to strike a new balance between resources and upcoming hazards (Ericksen et al. 2018).

Vulnerability and Adaptation Strategy

In Bangladesh, there are more than 200 laws and by-laws exist to tackle the challenges in relation to climate change and environmental issues. Several environment-related strategies and policies also exist. Government can realize that good public policy needs to be matched by investments to ensure implementation. In recent years, several big investments have been made by the government by its own or some extent supported by the private partner or development partners. Yet, investments in the Environment, Forestry and Climate Change (EFCC) sectors have suffered from a lack of coherence and delivered uncertain results in terms of their overall impact.

Conclusion

Climate change is inevitable. Climate change will make the people of Bangladesh more vulnerable to disease like cholera, dengue, cardiovascular and respiratory disease, allergic disease and malnutrition. The preparation of Bangladesh to face the challenge of global warming is not enough and cannot be overlooked. In order to tackle the health and socio-economic effects, relevant stakeholders including policy-makers, program designers, program implementers, civil servants and civil society members need to have better understanding about both climate change and its possible impacts. Appropriate and relevant policy needs to be formulated and follow-up for proper implementation. In recent past, Bangladesh has many success stories especially in health sector and human development index to share with the rest of the world. Through proper understanding of climate change issue and its impending effects on health, timely initiation of preparedness activities with thorough participation of community, Bangladesh has the potential to be successful once again in combating health effects of climate change. For that, all stakeholders must act with proper implementation of the planned activities with full cooperation of community and health workforces at all levels.

References

Ahmed S, Islam MA, Haque R et al (2008) Malaria baseline socioeconomic and prevalence survey 2007. BRAC-RED 10:32–33. www.bracresearch.org/.../Malaria_Baseline_Survey_2007..pdf

Barnes CS (2018) Impact of climate change on pollen and respiratory disease. Curr Allergy Asthma Rep 18(11):59

Bern C, Chowdhury R (2006) The epidemiology of visceral leishmaniasis in Bangladesh: prospects for improved control. Indian J Med Res 123(3):275–288

Borroto RJ (1998) Global warming, rising sea level, and growing risk of cholera incidence: a review of the literature and evidence. Geo J 44(2):111–120

Choudhury Zamil MAH, Banu S, Islam AM (2008) Forecasting dengue incidence in Dhaka, Bangladesh: a time series analysis. WHO Regional Office for South-East Asia. http://www.who.int/iris/handle/10665/170465

Elahi F, Khan NI (2015) A study on the effects of global warming in Bangladesh. Int J Environ Monitor Anal 3(3):118–122

Ericksen NJ, Ahmed QK, Chowdhury AR (2018) Briefing document no 4. http://citeseerx.ist.psu.edu/viewdoc/download?doi=10.1.1.452.8302&rep=rep1&type=pdf. Accessed on 6 Oct 2018

Gilman E, Ellison J, Coleman R (2007) Assessment of mangrove response to projected relative sea-level rise and recent historical reconstruction of shoreline position. Environ Monit Assess 124(1–3):105–130

Grecequet M, DeWaard J, Hellmann JJ, Abel GJ (2017) Climate vulnerability and human migration in global perspective. Sustainability 9(5):720

Hasib E, Chathoth P (2016) Health impact of climate change in Bangladesh: a summary. Curr Urban Stud 4:1–8

IPCC (2007) Climate change 2007: impacts, adaptation, and vulnerability. Contribution of working group II to the fourth assessment report of the intergovernmental panel on climate change. In: Parry ML, Canziani OF, Palutikof JP, van der Linden PJ, Hanson CE (eds) Cambridge University Press, Cambridge, United Kingdom, p 1000

Kabir R, Hafiz TAK, Emma B, Caldwell K (2014) Climate change and public health situations in the coastal areas of Bangladesh. Int J Soc Sci Stud 2(3):109–116

Kabir I, Rahman MB, Smith W, Lusha MAF, Milton Ah (2016) Climate change and health in Bangladesh: a baseline cross-sectional survey. Glob Health Action 9:29609

Kjellstrom T, Freyberg C, Lemke B, Otto M, Briggs D (2018) Estimating population heat exposure and impacts on working people in conjunction with climate change. Int J Biometeorol 62(3):291–306

Mahmud I, Mbuya N (2015) Water, sanitation, hygiene, and nutrition in Bangladesh: can building toilets affect children's growth? A World Bank study. https://openknowledge.worldbank.org/bitstream/handle/10986/22800/9781464806988.pdf

Menne B, Bertollini R (2000) The health impacts of desertification and drought. Down to Earth 14:4–6

Noé A, Zaman SI, Rahman M, Saha AK, Aktaruzzaman MM, Maude RJ (2018) Mapping the stability of malaria hotspots in Bangladesh from 2013 to 2016. Malar J 17:259

Phung D, Nguyen HX, Nguyen HLT, Luong AM, Do CM, Tran QD, Chu C (2018) The effects of socioecological factors on variation of communicable diseases: a multiple-disease study at the national scale of Vietnam. PLoS One. 13(3):e0193246

Rahman A (2008) Climate change and its impact on health in Bangladesh. Regional Health Forum 12(1):16–26
Rahman A (2008) Climate change and its impact on health in Bangladesh. Regional Health Forum 12(1)
Sellers S, Ebi KL (2017) Climate change and health under the shared socioeconomic pathway framework. Int J Environ Res Public Health 15(1):E3
St. Louis ME, Hess JJ (2008) Climate change: impacts on and implications for global health. Am J Prev Medicine 35:527–538
Waits A, Emelyanova A, Oksanen A, Abass K, Rautio A (2018) Human infectious diseases and the changing climate in the Arctic. Environ Int 121(Pt 1):703–713
WHO climate and health country profile (2015) Bangladesh
World malaria report (2010) WHO global malaria programme. http://www.who.int/malaria/world_malaria_report_2010/worldmalariareport2010.pdf

Chapter 6
Climatic and Environmental Challenges of Tea Cultivation at Sylhet Area in Bangladesh

Md. Nazrul Islam, Sahanaj Tamanna, Md. Mizanur Rahman, Mohammad Ahmmed Ali, and Imran Mia

Abstract Tea industry of Bangladesh is one of the most important source of income. Sylhet, the northeastern divisional city of Bangladesh, is the major tea-producing region of the country. For this reason, the study area was selected in Sylhet region to find out the causes of fluctuation of recent tea production in the study area The yield of tea is greatly influenced by climatic parameter of a region, especially temperature, rainfall, humidity, soil P^H. The data were collected from the Bangladesh Meteorological Department (BMD) and tea production; rainfall data were collected from the tea estates and plantation, and other time series data were explored from the web site of Bangladesh tea board, the different statistical year-book of Bangladesh, BBS. SWOT analysis and croup model were used in this research. Scanty rainfall causes irreparable losses because irrigation is seldom used on tea plantations. On the other hand, heavy rains erode top soil and wash away fertilizers and other useful chemicals. An analysis of the results of field experiments with weather data shows that heavy or scanty or delayed rainfall adversely affected the growth and yield of tea. It is observed that tea leaf production is slightly increased with increase in total annual rainfall, but tea cannot tolerate stagnant water and waterlogged lowland. The temperature and humidity have no direct effect

Md. N. Islam (✉)
Department of Geography and Environment, Jahangirnagar University, Savar, Dhaka 1342, Bangladesh
e-mail: nazrul_geo@juniv.edu

S. Tamanna
Bangladesh Environmental Modeling Alliance (BEMA), A Nonprofit Research and Training for Society Welfare, Mirpur-01, Dhaka, Bangladesh

Md. M. Rahman
Bangladesh Public Administration Training Centre, Savar, Dhaka 1343, Bangladesh

M. A. Ali
Ministry of Public Administration, Dhaka, Bangladesh

I. Mia
Department of Geography and Environment, Jahangirnagar University, Savar, Dhaka 1342, Bangladesh

© Springer Nature Switzerland AG 2021
Md. Jakariya and Md. N. Islam (eds.), *Climate Change in Bangladesh*,
Springer Climate, https://doi.org/10.1007/978-3-030-75825-7_6

on tea leaf production. It may largely depend on soil quality and solar radiation. Tea shrubs grow better when shielded from strong sunlight or violent winds.

Keywords Fluctuation · Climatic parameter · Rainfall · SWOT · Croup model · Tea production · Bangladesh

Introduction

Tea is one of the most important non-alcoholic beverage drinks worldwide and has been gaining further popularity as an important "health drink" in view of its purported medicinal value (Carr and Stephens 1992; Chomchalow 1996; Cheruiyot et al. 2010; Mamun 2011a, b; Mamun and Ahmed 2011; Ali et al. 2014; Ahmed and Ahmed 2015; Ali et al. 2015; Rahman et al. 2017). It is served as morning drink for nearly 2/3rd of the world population daily (Sumi 2020). The Bangladesh tea industry is one of the major sources of income for the national exchequer. Presently, this industry is facing a multitude of problems (Mondal et al. 2004; Chowdhury and Ward 2004; Muthaiya et al. 2013; Li et al. 2015). Lack of capital and modern machinery, lower market value of made tea in comparison to increasing production cost, lower yield per hectare in comparison to increasing domestic need and lack of modern techniques for measuring quality of tea constitute some of the nagging problems. There is also lack of perennial water source for irrigation during dry season or during prolonged drought (Islam et al. 2005; Ahmmed 2012; Ahmad and Hossain 2013; Hassan 2014; Hossain 2015). In addition, some owners of the tea gardens are not using government. Malnutrition among the children of the labor line, security problems of the executives, deterioration of law and order situation of the tea estates (log stealing, political or outsider influence on their internal arrangements, illegal occupation of land by the outsiders), lack of medical facilities for labor and lack of infrastructure (road, quarter, water supply network, etc.) are some of the other constraints (BCAS 1994; BBS 2000; Seran et al 1999; Bekhit 2006; Yang et al. 2006).

For successful tea culture, the above problems facing both the manufacturing and the marketing sector need to be addressed immediately. In Bangladesh, there is thus dire need to focus attention on improvements in the manufacturing sector covering quality of tea, its productivity, cost of production as well as the marketing system (Islam et al. 2005; Ahmmed 2012; Ahmad and Hossain 2013; Hassan 2014; Hossain 2015). The government tried to revive the sector in the early 1980s by privatizing and rehabilitating two tea estates, which had been nationalized in the 1970s; restructuring the Tea Board; privatizing the six state tea factories; and revamping public research on tea (BCAS 1994; Seran et al. 1999; Bekhit 2006; Yang et al. 2006). These policy initiatives have had some success, but much remains to be done to fully revitalize the tea sector. Infrastructure is still inadequate. The tax system is too complex, with too many taxes and rates that are too high. Despite the restructuring, the Tea Board and the Ministry of Agriculture are still too powerful (Sing et al. 2012). And trade policy needs to be revised to allow imports of made tea and exports of green leaf. The tea industry finds itself facing many challenges that could change the geography and economy of the landscape (Banerjee 1992; Alam 2002; Ahmed 2014).

More than any other crop, tea plantations have changed the face of many countries, but now they face the threats of climate change, the effects of deforestation (from when the original forests were being replaced by the tea plant, water shortages, and demands of fair wages from workers (Islam et al. 2005; Ahmmed 2012; Ahmad and Hossain 2013; Hassan 2014; Hossain 2015). The increase in average temperatures can affect the production for tea industry. As for low compensation among workers to identify a minimum wage to be allocated to farmers who work on tea plantations, as sometimes worker's earnings do not even reach the threshold necessary to be considered a living wage (BCAS 1994; Seran et al. 1999; Bekhit 2006; Yang et al. 2006). Bangladesh is an important tea-producing country. It is the 12th largest tea producer in the world. Its tea industry dates back to British rule, when the East India Company initiated the tea trade in Chittagong in 1840. Today, the country has 166 commercial tea estates, including many of the world's largest working plantations. The industry accounts for 3% of global tea production and employs more than 4 million people (Mamun 2011a, b; Mamun and Ahmed 2011; Nasir and Shamsuddhoa 2011; Haque 2013; Ali et al. 2014; Ahmed and Ahmed 2015; Ali et al. 2015; Rahman et al. 2017). The tea is grown in the northern and eastern districts, the highlands, temperate climate, humidity, and heavy rainfall within these districts provide a favorable ground for the production of high-quality Bangladesh Tea Industry established at 1840 when a pioneer tea garden was established on the slopes of the hills in Chittagong where the Chittagong Club now stands. First commercial tea garden was established in 1857 at Mulnichera in Sylhet (Bangladesh Tea Board (BTB) 2009; BTD 2012; Hossain et al. 2013; Rahman 2016; BTB 2017).

To retain these consumers, tea growers shifted to the tea bag culture from loose tea. But customers took further steps forward, driven by health consciousness and economic development. In the past, people used to drink normal black tea (Mamun 2011a, b; Mamun and Ahmed 2011; Ali et al. 2014; Ahmed and Ahmed 2015; Ali et al. 2015; Rahman et al. 2017). But now, they are not satisfied with the normal tea and moved to organic and value-added products, concerned about their health benefits and taste. Every year, 3–5% of people tend to consume from the premium segment in general (Islam et al. 2005; ITC 2015). Keeping that in mind, consumer products manufacturers have to produce premium products. Tea is not outside this trend (BCAS 1994; Seran et al. 1999; Bekhit 2006; Yang et al. 2006). Tea consumption is an indicator of national economic growth. More consumption of tea means people have enough money in their pockets," Nader Khan, chairman of Halda Valley, a producer of premium tea. Introducing new products and gaining acceptance is not an easy task. Though the sector has enormous potential to grow in local and foreign markets, it faces numerous challenges (BCAS 1994; Seran et al. 1999; BTRI 2003; Bekhit 2006; Yang et al. 2006). To retain local consumers and capture export markets, Bangladesh has to increase production capacity.

According to Bangladesh Tea Board data, in 2017, Bangladesh produced 79 million kg tea against a consumption of nearly 86 million kg. Those involved with the industry called for finding new areas for cultivation and clearing more land in Chittagong and Sylhet (Islam et al. 2005; Ahmmed 2012; Ahmad and Hossain 2013; Hassan 2014; Hossain 2015). There is a common perception that tea cultivation is not possible without hilly areas, which is not always true. Tea garden in Panchagarh is the proof, they said. "Exploring new markets is a big challenge since it is occupied by other countries. Maintaining the premium quality is the key to success in capturing foreign markets. Tea is a regular export item of Bangladesh. But due to slow growth of production growing consumption and stiff competition from other tea-exporting courtiers, tea export of the country has declined (BCAS 1994; Seran et al. 1999; Bekhit 2006; Yang et al. 2006; BTRI 2012). Export of tea to different countries by Bangladesh witnessed substantial changes over time. In order to cope with such changes, it is necessary to look for new buyers and pursue an aggressive policy in the world tea market. Domestic consumption of tea has steadily increased. This trend is likely to remain and even may gain momentum. In the face of rising domestic consumption, a stable level of tea export can be maintained only by a sustained increase in tea production. Various aspects of tea production and export have been dealt with, among others, by Khalid, and Sabur (2002).

Rationale of the Study

Tea is one of the most important plantation crops, with a financially viable lifespan of not less than 60 years; the Sylhet tea industry has a market. Interestingly, the majority of tea produced in Sylhet is consumed domestically. For example, around 80% was consumed domestically, indicating the great domestic demand for this product. Nevertheless, it also is a major foreign currency earning commodity for most tea-producing countries (Mondal et al. 2004; Chowdhury and Ward 2004; Muthaiya et al. 2013; Li et al. 2015). That gives the tea industry a noteworthy place in the country's economy as an earner of foreign currency. In addition, tea planting provides lucrative employment to a large number of both poorly educated people and well-educated ones. It is labor-intensive and offers employment opportunities to both female and male workers. Tea plantation offers work to underprivileged people living in the undeveloped regions and creates employment opportunity per household to these people (Islam et al. 2005; Ahmmed 2012; Ahmad and Hossain 2013; Hassan 2014; Hossain 2015). Apart from that, the tea industry supports the growth of several ancillary industries. It is one of the major stakeholders in the jute and plywood industries, as tea is generally packed in either jute bags or plywood chests for transportation. A huge amount of coal or gas is required for manufacturing made tea, which helps the coal and petroleum industries. Additionally, it is a major user of fertilizer, pesticides, and weedicides.

Ecotourism is another important aspect of preserving and sustaining the diversity of the natural and cultural environments, and it has ushered in a novel additional method of income for tea estate inhabitants (BCAS 1994; Seran et al. 1999; Bekhit 2006; Yang et al. 2006). Visitors to the estates are put up in a way that is minimally intrusive and sustains the native cultures and local products, and thus, tea growers can expose their products to a large audience (Islam et al. 2005; Ahmmed 2012; Ahmad and Hossain 2013; Hassan 2014; Hossain 2015). Moreover, the revenue generated by ecotourism encourages local bodies to fund developmental projects to preserve the flora, fauna, and cultural heritage of their area. Thus, apart from soil erosion, the water table is lowered, with consequent harmful effects on the moisture status of the surface layers of soil, compelling the tea plants to literally withstand lingering periods of drought (Mondal et al. 2004; Chowdhury and Ward 2004; Muthaiya et al. 2013; Li et al. 2015). Alternatively, decades of studies formulated the application of inorganic fertilizers for gearing up productivity. However, due to the poor organic status of the soil and deterioration of water-holding capacity, infiltration is retarded, with consequential runoff problems. The problem has been aggravated by the continual broadcast of ammonium sulfate, which led to a lowering of ph (Mamun 2011a, b; Mamun and Ahmed 2011; Ali et al. 2014; Ahmed and Ahmed 2015; Ali et al. 2015; Rahman et al. 2017). The recurring use of nitrogenous fertilizers causes the release of other elements from bound sites and leaching out into rivers and other water bodies, rendering them absolutely irreconcilable with biodiversity.

In some tea-growing countries, profuse and erroneous application of insecticides has disturbed the environment. As tea plantations are basically a monoculture, they offer the perfect conditions for a number of pests (Islam et al. 2005; Ahmmed 2012; Ahmad and Hossain 2013; Hassan 2014; Hossain 2015). In order to keep productivity to the desired level, a number of lethal, detrimental pesticides are extensively used. However, erroneous application of these substances obliterates beneficial predators and destroys biodiversity as well. Insecticides are dispersed by fossil fuel-driven power sprayers, which produces mist that is harmful to the local environment, especially human beings (BCAS 1994; Seran et al. 1999; Bekhit 2006; Yang et al. 2006; Cai et al. 2013; Butt et al. 2015). Further, if not applied well ahead of plucking, pesticides with higher half-lives may produce residual effects on liquor that can put human life in jeopardy. Alternatively, though broad-spectrum insecticides provide many benefits by means of good control, higher yields, and handsome profits, they also have raised grave concerns, such as growth of resistance against pesticides, the reappearance of pests, an epidemic of secondary pests, detrimental effects to the environment, and unwanted residual effects in the liquor (Gurusubramanian et al. 2008; Mamun 2011a, b; Mamun and Ahmed 2011; Ali et al. 2014; Ahmed and Ahmed 2015; Ali et al. 2015; Rahman et al. 2017).

Origin and Spread of Tea Cultivation

Tea was used as a medicinal beverage for the first time in China. According to the *Ben Chao*, the Chinese medical book, the role of tea in Chinese history can be dated to the year 2737 BC, when the emperor, Shen Nong, discovered it (Hara et al. 1995;

Harbowy et al. 1997). The people of Southwest China used teas for paying tribute to the Chinese emperors (Gamborg et al. 1968; Forrest 1969; Dubravina et al. 2005; Friedman 2007) . A Chinese landlord, Wang Bao, wrote an essay called "Tong Yue" in 59 BC, in which the making and sale of tea was mentioned. In addition, excavated tombs dating to 200 BC revealed that tea used to be included with the burial objects (Hara et al. 1995). "Cha Jing," written by Lu Yu in 780 ad, highlighted the fact that, tea was a staple commodity in China (BCAS 1994; Seran et al. 1999; Bekhit 2006; Yang et al. 2006). The book highlighted the origin, characteristics, names, and qualities of tea, along with details about tea plantations, plucking, processing, making, drinking, and storage (Chow and Kramer 1990). Southeast China, near the source of the Irrawaddy River, is considered to be the probable center of origin of tea (Taylor and Mcdowell 1991). Later, it was taken to South parts of China and beyond, to some parts of Southeast Asia. From there, by 221 BC, tea spread to a number of tropical and subtropical countries (Hara et al. 1995) by migration of peoples in war time, including Vietnam, Myanmar, Laos, and Thailand.

Hence, techniques of tea processing in certain mountainous regions of these countries are somewhat similar to those employed in ancient China (Zhuang 1988). During the fifth century, China established the tea trade with Turkey, which further extended to Iran, Rome, Arabian Peninsula, Afghanistan, Pakistan, and Korea, creating the Silk Road. Zen Buddhist missionaries introduced tea to Japan in 805 AD as a medicine due to its meditation-enhancing properties (BCAS 1994; Seran et al. 1999; Bekhit 2006; Yang et al. 2006). There are two schools of thoughts regarding the invasion of tea to Europe. It had been suggested that a Portuguese Jesuit father first introduced tea to Europe in 1560. However, other studies suggested that the first tea reached Europe around 1610, carried by Dutch ships from Java (Chow and Kramer 1990). The first cultivations introduced in various countries were in Indonesia in 1684, Russia in 1833, Sri Lanka in 1839, Malawi in 1875, Iran in 1900, Kenya in 1903, and Argentina in 1924 (Chen and Cheung 1987).

Tea Cultivation History in This Region

The art of tea cultivation in Bangladesh began over a century and a half ago in the 1840s near the Chittagong Club. The first tea garden to be established was Malnicherra in Sylhet in 1854. Its commercial production began shortly thereafter in 1857 (Nasir and Shamsuddhoa 2011). Today, the main tea-growing areas lie to the east of the Ganga-Jumma flood plain in the hill areas bordering India's Cachar tea-growing district. Most of Bangladesh tea grows at only 80–300 ft. above sea level northeast of Sylhet in the Seven Valleys (Islam et al. 2005; Ahmmed 2012; Ahmad and Hossain 2013; Hassan 2014; Hossain 2015). Tea is still grown in Chittagong as well as in the Hill Tracts. During pre-partition days and up to the year 1947, all teas produced in the Sylhet and Surmah valleys were called Indian teas, but were also known as Surmah valley teas. Crop figures for the region during the mid-1940s were approximately 10–15 million kg per year (Mamun 2011a, b;

Mamun and Ahmed 2011; Ali et al. 2014; Ahmed and Ahmed 2015; Ali et al. 2015; Rahman et al. 2017). The teas were all of Orthodox manufacture, their quality being fairly similar to that of neighboring Cachar district teas. There was also some Legg-Cut and green tea manufacture prior to 1947 (*Source: Salman Ispahani, Member, Bangladeshiya Cha Sangsad*).

Tea Cultivation History Post-Partition 1947–1971

After partition, the subcontinent was broadly divided into two political regions India and Pakistan (comprising West and East Pakistan). When Pakistan became independent in 1947, there were 133 tea gardens. By 1971, this number had risen to 147, with roughly 90,000 workers out of a total country population of 249,000 people (Mamun 2011a, b; Mamun and Ahmed 2011; Ali et al. 2014; Ahmed and Ahmed 2015; Ali et al. 2015; Rahman et al. 2017). In 1950, under the Pakistan Tea Act, the Pakistan Tea Board was established in Dhaka, and in 1957, the Tea Research Institute was founded in Srimangal in 1957. Together, these organizations aimed at promoting the sale and consumption of tea in Pakistan and abroad, and at assisting in the research and development of the tea industry (Khisa and Iqbal 2001; Mehrin et al. 2016; Kfoury et al. 2018). The Tea Ordinance Act of 1959 replaced the earlier Pakistan Tea Act of 1950 to enhance the Board's role in promoting tea cultivation and quality control (BCAS 1994; Seran et al. 1999; Bekhit 2006; Yang et al. 2006). During the 1952–1953 seasons, buyers, sellers, and brokers in Chittagong got together, under the auspices of the Pakistan Tea Association, to form the Tea Traders Association of Chittagong. This association's duty was to promote the common interests of tea sellers and buyers in the Chittagong market. In 1960, the Tea Traders Association of Pakistan was registered (Khisa and Iqbal 2001).

However, by the late 1960s, the need for better quality teas was strong and, with the gradual decline in the availability of Orthodox teas, CTC teas, particularly the better liquoring types, received strong support (Mondal et al. 2004; Chowdhury and Ward 2004; Muthaiya et al. 2013; Li et al. 2015). It was noticeable during this period that the consumer gradually demanded brighter and better teas, and at the same time, became partial to the strong liquor produced by the CTC and the Legg-Cut methods of manufacture. Until 1971, teas continued to be imported to meet the growing internal demand while production was inadequate. With export restrictions and the captive market of West Pakistan, the tea industry operated in a seller's market with the 1969 crop being sold in the Chittagong auctions at Rs 3.50 per lb., against the CIF Chittagong cost of imported tea at Rs 1.50 per lb (Islam et al. 2005; Ahmmed 2012; Ahmad and Hossain 2013; Hassan 2014; Hossain 2015).

Tea Cultivation History During the 1971 War of Liberation

During the War of Liberation in 1971, the tea industry suffered many setbacks. In addition to the fact that many factories were damaged, two-thirds of the experienced planters of British and Pakistan origin left the industry, while many senior Bangladeshi planters joined the war. This meant that inexperienced men who had to combat disturbed conditions were running the estates (Islam et al. 2005; Ahmmed 2012; Ahmad and Hossain 2013; Hassan 2014; Hossain 2015). Some of the battles even took place in the tea garden areas, which were very near the borders. In fact, for the first time after the crackdown on April 4, 1971, the available senior offices of the Eastern Sector of the Liberation War met at Teliapara Tea estate manager's bungalow, which became a seat of the Bangladesh Forces Headquarters (BDFHQ) for quite a time. In 1947, gardens in East Pakistan had 75,000 acres (30,364 ha) under tea (Sharma et al. 1981; Prakash et al. 1999; Singh et al. 2006; Saha 2010; Smith 2012). Between 1947 and 1960, the acreage devoted to tea increased by 8000 acres. In the following decade, thanks to the compulsory 3% extension plan undertaken by the government (Mamun 2011a, b; Mamun and Ahmed 2011; Ali et al. 2014; Ahmed and Ahmed 2015; Ali et al. 2015; Rahman et al. 2017). In 1947, Pakistan began with a production of 41.5 million lbs. (19.0 million kg)—approximately 7% of India's production. In 1956, this figure reached 53 million lbs. (24.1 million kg), and in 1971, a record crop of 69.18 million lbs.

Tea Cultivation History After 1971 War of Liberation

After the war, assistance from England was readily available. At the request of the Government of Bangladesh, the British agency Overseas Development Administration (ODA) commissioned the Commonwealth Development Corporation (CDC) in 1973 to assess the requirements for a process of rehabilitation and reorganization of the tea industry, including tea growing, manufacture, research, markets, and market organization, with an assessment of the financial and economic returns to such a program (Mondal et al. 2004; Chowdhury and Ward 2004; Muthaiya et al. 2013; Li et al. 2015). In 1977, the Bangladesh Tea Board was reconstituted with objectives common to those of the erstwhile Pakistan Tea Board formed under the Pakistan Tea Act 1950, and as the regulatory body for the tea industry of Bangladesh, the role of the Tea Board expanded to include the monitoring of the crop and its disposal, the issuance of export licenses to export buyers, and the authority to give permission to producers for consignment and direct sales, etc. (BCAS 1994; Seran et al. 1999; Bekhit 2006; Yang et al. 2006; Mamun 2011a, b; Mamun and Ahmed 2011; Ali et al. 2014; Ahmed and Ahmed 2015; Ali et al. 2015; Rahman et al. 2017). In 1974, the Tea Traders Association of Bangladesh replaced the Tea Traders Association of Pakistan. Eight years later, the metric weight system was adopted for the sale of tea, replacing the earlier imperial system.

The area that produced tea increased from around 43,000 ha in 1971 to the present area of about 48,000 ha. After 1971, an improvement in yield per hectare was also evident.

By the 1973–74 season, the production of Bangladesh tea recovered to pre-1971 levels of around 30–32 million kg. During 1975–76, in an attempt to increase yields, the Tea Board prepared two plans for intensive cultivation and replanting. Bangladesh Tea Research Institute BTRI, the scientific wing of the Tea Board, also brought out several high yielding clonal varieties of distinct character and quality (Mondal et al. 2004; Chowdhury and Ward 2004; Muthaiya et al. 2013; Li et al. 2015). By 1979, British consultants had developed a strategy to rehabilitate the damaged tea industry of Bangladesh. Although by this time an onward program for intensive cultivation and replanting of the tea was going on, the actual thrust started in 1983–84 and was effective from 1985/85 (Mamun 2011a, b; Mamun and Ahmed 2011; Ali et al. 2014; Ahmed and Ahmed 2015; Ali et al. 2015; Rahman et al. 2017). During the years 1971–1994, production increased from 24.2 million kg (53.2 million lbs.) in 1972 to 52.1 million kg (115 million lb) in 1994. Although the number of estates manufacturing Orthodox teas was declining (especially after the 1980s), after independence, Orthodox teas continued to form the bulk of all teas produced.

Aims and Objectives of This Study

The aim of my research is to find out the causes of fluctuation of recent tea production in the study area. The specific objectives of the research are:

1. To study the present environmental conditions for tea cultivation,
2. To analyze the environmental challenges of tea cultivation in Bangladesh,
3. To propose a crop model for developing a sustainable tea cultivation.

Research Questions

1 How the present tea cultivation systems are facing the problems to cope with the climate change?
2 Does any challenges of tea cultivation systems in Bangladesh? If yes, what are they?
3 How can we decrease the current environmental challenges for tea cultivation in Bangladesh?
4 Do the propose model could be the helpful to mitigate the barriers of tea cultivation in Bangladesh?

Data Sources and Methodology

Sylhet district is the major tea leaves producing zones along with other agricultural crops due to its climatic condition and geographic position. Therefore, this area was selected purposively to conduct this study. An exploratory survey on collect information on the fluctuation pattern of tea leaf production and changing impacts of climatic parameters (rainfall, temperature and humidity) on tea leaf yield (BCAS 1994; Seran et al. 1999; Bekhit 2006; Yang et al. 2006; Mamun 2011a, b; Mamun and Ahmed 2011; Ali et al. 2014; Ahmed and Ahmed 2015; Ali et al. 2015; Rahman et al. 2017). The climatic data of Sylhet district were collected from the Bangladesh Meteorological Department (BMD); and tea production, rainfall data were collected from the selected tea estates and plantation, and other time series data were explored from the web site of Bangladesh tea board, the different statistical year book of Bangladesh, BBS. The published data of tea estates were collected from their administrative office and from the Banglapedia. This study was carried out over a period of six month ranging from January 2019 to March 2019. The data were analyzed graphically by using MS Excel, SPSS.

Temperature and Precipitation Data

Monthly tea yield data were collected from 32 tea gardens across study area with observations from January to March 2019. The tea yield data were generated for each plucking round which follows a seven to ten day cycle (BCAS 1994; Seran et al. 1999; Bekhit 2006; Yang et al. 2006; Mamun 2011a, b; Mamun and Ahmed 2011; Ali et al. 2014; Ahmed and Ahmed 2015; Ali et al. 2015; Rahman et al. 2017). The green leaf plucked by laborers from the pruned and unpruned sections are weighed separately and summed for the entire month. The data from each round is recorded on a daily basis by garden managers. From these records, tea yield can be calculated as the total green leaf weight (kg) divided by the area of the plantation (ha). The temperature data used in this study were obtained from the Bangladesh Meteorological Department (BMD) gridded datasets (Mondal et al. 2004; Chowdhury and Ward 2004; Muthaiya et al. 2013; Li et al. 2015).The BMD gridded data include a daily temperature product at 1° spatial resolution. These datasets were derived from ground-based station data interpolated using a modified version of Shepherd's angular distance weighting algorithm (Sharma et al. 1981; Prakash et al. 1999; Singh et al. 2006; Saha 2010; Smith 2012). We used the Precipitation Estimation from Remote Sensing Information to measure precipitation.

Experimental Design, Materials and Methods

These were used to acquire the summary data provided in this article from tea producers. Questionnaire was used for plantation managers and another used from smallholders following data collection, processing was undertaken to summarize findings into tables and figures which summarized the data by each of the tea-growing regions (BCAS 1994; Seran et al. 1999; Bekhit 2006; Yang et al. 2006; Mamun 2011a, b; Mamun and Ahmed 2011; Ali et al. 2014; Ahmed and Ahmed 2015; Ali et al. 2015; Rahman et al. 2017). Questionnaire categories for fertilizer application were devised in relation to the Tea Research Association industry standards to enable analysis of data as to where standards were being followed. In total, 22 plantation managers and 17 smallholders completed questionnaires for use in this research. Participants were identified via response to invitations sent to Tea Research Association member gardens to attend one of the workshops where questionnaires were undertaken (Islam et al. 2005; Ahmmed 2012; Ahmad and Hossain 2013; Hassan 2014; Hossain 2015). Entries were checked for input errors and outliers by my honorable supervisor. Where outliers were likely a respondent error by the participant, these particular question answers were excluded from the analysis. Participants volunteered their time to participate in the workshops and were provided with light food and drink refreshments.

There are as many as 18 tea gardens existing in Sylhet Sadar Upazila and Jaintapur Upazila. Out of 18 tea gardens, eight were selected by random sampling method considering the size and location of each garden. The tea gardens that were selected for the study were Malnichhera, Lackaturah, Star tea estate, Alibahar, Doldoli tea estate, Khadim tea estate, Burjan tea estate of Sylhet Sadar Upazila and Habibnagar tea estate of Jaintapur Upazila (BCAS 1994; Seran et al. 1999; Bekhit 2006; Yang et al. 2006). The proportionate number of households and the smooth communication system of each garden have been taken into consideration while selecting the total number of respondents. In order to ensure proper representation, respondents were selected from five broad categories of managements, namely Sterling Companies, National Tea Companies, Bangladesh Tea Board, Private Company Limited and proprietary. A latest list of the number of households existing in each tea garden was collected from the individual tea garden for the purpose of selecting the household respondents (Mamun 2011a, b; Mamun and Ahmed 2011; Ali et al. 2014; Ahmed and Ahmed 2015; Ali et al. 2015; Rahman et al. 2017). Interview schedule and observation methods were mostly used to collect requisite data for the study. At first, a draft schedule was prepared considering the objectives of the study. Then, the schedule was pretested in the selected area among a few tea garden workers. After making necessary modification and correction, the final schedule was prepared. The interview schedule was both close and open ended (Mondal et al. 2004; Chowdhury and Ward 2004; Muthaiya et al. 2013; Li et al. 2015). The research assistants collected data following survey method, using interview schedule. The data are collected during the period of the months January to March 2019. After the collection of whole range of data, they were processed and tabulated. Editing, coding, and decoding of collected data were also done simultaneously, avoiding irrelevant and unreliable information. The tabulated data were analyzed and described according to the aims and objectives of the study, using simple statistical techniques.

Four types of interview schedule were prepared (Mamun 2011a, b; Mamun and Ahmed 2011; Ali et al. 2014; Ahmed and Ahmed 2015; Ali et al. 2015; Rahman et al. 2017). The total member of randomly selected respondents was 1216. Out of 1216 respondents, the number of male, female, and adolescent boys/girl's respondents was 1190. Out of 1190 respondents, 112 respondents were adolescent boys/girls. There were 18 panchayat members, 06 staffs of management committee of each tea garden, and 02 union council members among the total member of 1216 respondents. The age range of the respondents was (15–70) years.

Integrating SWOT and Crop Suitability (CS) Model

The methodology for the definition of the crop/land suitability map for a given area is based on the processing and combination of a set of layers that are considered as the main drivers for land suitability:

- Soil texture
- Physical properties (depth and texture)
- Chemical properties (organic matter (OM) and ph)
- Climatic conditions
- Precipitation
- Temperature
- Topography (slope)
- Current land use/cover

SWOT Analysis

A SWOT analysis is a great way to guide business-strategy meetings. It is powerful to discuss the core strengths and weaknesses and then move from there to define the opportunities and threats, and finally to brainstorming ideas. Oftentimes, the SWOT analysis envision before the session changes throughout to reflect factors were unaware of and would never have captured if not for the group's input. A SWOT for overall strategy sessions or for a specific segment. This way, one can see how the overall strategy developed from the SWOT analysis will filter down to the segments below before committing to it. Work in reverse with a segment-specific SWOT analysis that feeds into an overall SWOT analysis. An adaptation of the SWOT analysis is Weihrich's TOWS matrix. (BCAS 1994; Seran et al. 1999; Bekhit 2006; Yang et al. 2006). The matrix identifies potential tactical strategies that could be deployed for the purpose of exploiting opportunities or defending against threats through the leverage of the existing strengths and the reduction of weaknesses. The TOWS matrix seeks to develop tactical strategies based on four different positions.

	Weakness (W)	**Strengths (S)**
Opportunities (O)	Examines strategies that take advantage of opportunities to avoid weaknesses (WO)	Examines strategies that use strengths to make use of opportunities (SO)
Threats (T)	Examines strategies that minimize the effect of weaknesses and overcome or avoid threats (WT)	Examines strategies that use strengths to overcome or avoid threats (ST)

SWOT Analysis

Strength	Weakness	Opportunity	Threat
1. Experienced man power of BTRI and PDU. 2. Effective and Cooperative communication among Tea Board and producers. 3. Contribution of Agricultural Bank and Rakab in financial management. 4. High prospects (Internal and External) market	1. Fund limitation 2. Shortage of labor 3. Cultivation in slow mode 4. Unawareness in cultivation 5. Ownership conflict 6. Land lease no performing	1. Maximum land utilization 2. Expansion of small scale tea cultivation 3. Earning foreign currency 4. Employment creation 5. Increasing internal demand 6. Rural poverty elevation 7. Progressive production and quality of tea.	1. Environmental degradation for global warming 2. Tea import tendency 3. Increasingly export decrease 4. Lower obsessiveness of production 5. Expenditure of production

The WO strategy in the first quadrant attempts to maximize opportunities arising from the external environment and eliminating the internal weaknesses that hinder its growth. For example, an electronics manufacturer may be aware of the growing demand for tablets, but may lack the technology required for producing screens. Possible strategies would be to create joint ventures with firms that have the technology and possibly the patents for interactive screens, which are at the edge of innovation (Mondal et al. 2004; Chowdhury and Ward 2004; Muthaiya et al. 2013; Li et al. 2015). Another option would be to subcontract the function to firms that have competency in this field. If no action is taken, the opportunity of growth is left to competitors. The SO strategy in the second quadrant is an ideal situation where maximize on both strengths and opportunities. The ST strategy uses the organization's internal strengths that can counteract threats from the greater environment. The WT strategy in the fourth quadrant is the worst-case scenario when has to minimize both its weaknesses and its threats. However, external forces may not be avoidable as in the case of the tea industry.

Results and Discussion

Investment in Bangladesh tea is very low compared to national investment. The growths of turnover and investment in tea have been shown in Fig. 6.3. Bangladesh Tea has developed itself as an agro-based, labor-intensive, and export-oriented sector and plays an important role in the national economy through export earnings, trade balancing, import substitution, and employment generation (Mamun 2011a, b; Mamun and Ahmed 2011; Ali et al. 2014; Ahmed and Ahmed 2015; Ali et al. 2015; Rahman et al. 2017). Bangladesh tea dates back to 1854 when the first tea estate was established at Malnicherra in Sylhet. Presently, Bangladesh has 172 tea estates with grant area 116,264 ha of which 56,846 ha. (2011); i.e., 48.89% is under tea cultivation. Tea estates in Bangladesh are predominantly in the private sector. Managements are the only players for investments in tea estates (BCAS 1994; Seran et al. 1999; Bekhit 2006; Yang et al. 2006; Mamun 2011a, b; Mamun and Ahmed 2011; Ali et al. 2014; Ahmed and Ahmed 2015; Ali et al. 2015; Rahman et al. 2017). The private owners possess 97.54 percent of grant area and Bangladesh Tea Board owns remaining 2.46 percent of land, which is a statutory body under the Ministry of Commerce. Srimangal in Moulvibazar, called as the tea capital of Bangladesh, is the main center of tea area commonly known as Surma Valley. Greater Sylhet, the tea granary of Bangladesh, has 133 tea estates. Quality tea is also grown 23 tea estates in Chittagong and Chittagong Hill Tracts known as Halda Valley of Country's famous tea-growing areas. Tea cultivation has also started in 2000 in 16 tea estates in Panchagarh, the northern district of the country. Though tea industry suffered a serious setback in 1971, but Bangladesh could succeed in reversing with the help of the government, foreign assistance and hard work of planters. Per ha yield has increased from 500 kg per ha then to over 1240 kg per ha today (Mondal et al. 2004; Chowdhury and Ward 2004; Muthaiya et al. 2013; Li et al. 2015). The country is planning to increase its production to an average of over 1500 kg per ha in a few years' time. Bangladesh Tea Board has undertaken measures to improve the quality of tea by extending the area with new varieties of hybrid clones, modernizing factories and improving infrastructure. The annual production of tea is now 59.13 million Kg of made tea (2011). Investment in Bangladesh tea is very low compared to national investment. The growths of turnover and investment in tea have been shown in Fig. 6.3. While the turnover is increasing in tea, the investment is decreasing in Bangladesh tea.

Environmental Condition of Tea Cultivation and Processing System in Sylhet

Climatically, tea belongs to the monsoon lands where high temperatures, long growing season, and heavy rainfall help the growth of tea plants. A temperature of 21 °C during the growing season of not less than eight months is ideal (Mamun 2011a, b; Mamun and Ahmed 2011; Ali et al. 2014; Ahmed and Ahmed 2015; Ali et al. 2015; Rahman et al. 2017). Tea requires a moderately hot and humid climate. Climate influences yield, crop distribution, and quality. Therefore, before cultivating tea in a new area, the suitability of the climate is the first point to be considered. Tea grows best on well-drained fertile acid soil on high lands (BCAS 1994; Seran et al. 1999; Bekhit 2006; Yang et al. 2006). Effects of climate change on tea production. Climate and geography are key factors in determining both where tea can be grown and how the tea grown in a particular region tastes. It is first important to understand the basic growing requirements of the tea plant. The tea plant is highly adaptable, and can grow in a broad range of condition. Climate change brings both advantages and disadvantages for the growth and development of tea and will ultimately have a considerable impact on production (Mondal et al. 2004; Chowdhury and Ward 2004; Muthaiya et al. 2013; IPCC 2014; Li et al. 2015). The beneficial changes include increases in temperature and CO_2. The adverse impacts include a decrease in rainy days and in relative humidity, and an increase in climate extremes and variations, such as drought, flood, and extremely cold and hot weather. These adverse climate changes will cause serious problems for tea production and sustainable development. The impact of the reduction of sunny days will depend on the degree of change and the location.

Even in areas equally capable of growing the tea plant, the qualities of the finished tea can be profoundly influenced by climate conditions. This is caused both by changes in chemical composition of the tea plant in response to different growing condition and by chemical changes that occur during processing (Islam et al. 2005; Ahmmed 2012; Ahmad and Hossain 2013; Hassan 2014; Hossain 2015). For example, in cooler and drier air, common at high elevations, when processing tea as one would make black tea, it is possible to achieve a "hard wither," where the tea leaves dry out before they are fully allowed to oxidize. In hot, humid air, this same process is not possible (Fig. 6.1).

Due to climatic condition and geographical position, north-eastern region of Sylhet district is the major tea leaves producing zones along with other agricultural crops (BCAS 1994; Seran et al. 1999; Bekhit 2006; Yang et al. 2006; Mamun 2011a, b; Mamun and Ahmed 2011; Ali et al. 2014; Ahmed and Ahmed 2015; Ali et al. 2015; Rahman et al. 2017). The time series data of 2012–2017 on the yield,

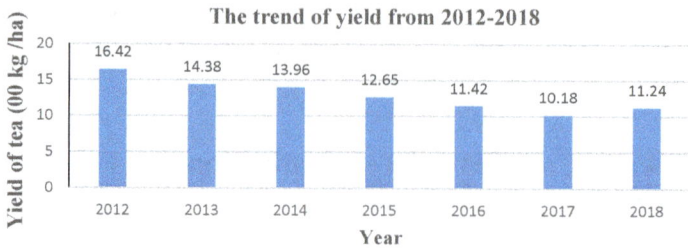

Fig. 6.1 Trend of the yield of tea in Sylhet from 2012 to 2018

rainfall, maximum temperature, and minimum temperature were collected from this tea estate. Simple linear regression was done to correlate yield, rainfall, and temperature. The trend of yield from 2012 to 2018.

Pattern of Tea Leaf Production in Selected Tea Estates

Production and quality of tea leaf in the selected four tea estates was different. The maximum average (for one year) tea leaf production was found in Loobacherra tea estate (1835.7013 kg/ha), and lowest was found in Lakkatoora (682.30811 kg/ha). Production of others two estates Burjan and malnichara was 877.8412 kg/ha and 1367.3418 kg/ha, respectively. This variation may be due to the soil property and management potentiality of these tea estates (Islam et al. 2005; Ahmmed 2012; Ahmad and Hossain 2013; Hassan 2014; Hossain 2015). In favor of the maintenance of good quality tea, it is required to collect one bud and two leaves (Hays et al. 2008), but in the field observation, it is noticed that in Loobacherra tea estate collected more than two leaves and a bud. Thus, the quantity may increase but the quality of tea may hamper (Fig. 6.2).

Fig. 6.2 Average month-wise yield (2012–2018)

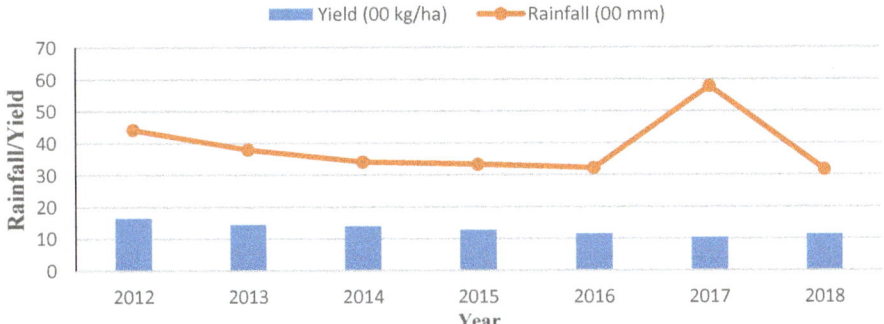

Fig. 6.3 Trend of yield and annual rainfall

The highest production was observed in the monsoon (June–October) followed by the summer (March–June) and winter (October–March). The monsoon dynamics, temperature, solar radiation, and precipitation played as key drivers for the highest production in the monsoon (Mamun 2011a, b; Mamun and Ahmed 2011; Ali et al. 2014; Ahmed and Ahmed 2015; Ali et al. 2015; Boehm et al. 2016; Rahman et al. 2017). The month of August produced the highest amount of leaves followed July and June. In January, production was lowest followed by February and March (BCAS 1994; Seran et al. 1999; Bekhit 2006; Yang et al. 2006). The year begins and ends with dry periods in Bangladesh. Tea zones experience a dry season from November to March while the rainy season continues from April to October and above 80% of annual rainfall is obtained during June–September (Mamun 2011a, b; Mamun and Ahmed 2011; Ali et al. 2014; Ahmed and Ahmed 2015; Ali et al. 2015; Rahman et al. 2017; Paul et al. 2017). From the recent Meteorological report of Bangladesh, it is evident that in April–May the temperature reaches up to the maximum of 40 °C, whereas during the December or January, the minimum temperature drops to the extent of 7.8 °C. The impact of climate change is already witnessed in the declining trend of tea production (Mondal et al. 2004; Chowdhury and Ward 2004; Muthaiya et al. 2013; Li et al. 2015). Most of the tea-producing regions are characterized by a specific wet and or a combination of alternate wet and dry season interspersed by the temperature fluctuations from mild to medium. Bhagat et al. (2018) reported that tea-producing countries like Sri Lanka, China, India, and Kenya have witnessed significant production loss due to climate change. It is projected that this change will be accelerated in future times. (Islam et al. 2005; Ahmmed 2012; Ahmad and Hossain 2013; Hassan 2014; Hossain 2015). The climate change not only change temperature, precipitation, relative humidity rainy days, and annual sunshine times, it also affect other basic elements of tea growth and development such as soil PH, moisture content soil organic matter and nutrients availability, pest and disease management, and ecological system around tea gardens and eventually the production and economy of tea.

Effects of Microclimatic Parameter on Tea Leaf Production

Microclimate has large impact on tea leaf production. The seasonal monsoon also greatly affects the tea leaf production (Anon 1996; Hicks 2001). In the selected tea estates, there was significant relationship with microclimates and tea leaf production (BCAS 1994; Seran et al. 1999; Bekhit 2006; Yang et al. 2006). More than the total amount, the distribution of rainfall matters a lot for sustained high yield of tea throughout the season. In the Sylhet region, the rainfall distribution is not even. The excess rainfall in the monsoon months of June–September causes drainage problems (Mondal et al. 2004; Chowdhury and Ward 2004; Muthaiya et al. 2013; Li et al. 2015). The average monthly rainfall during November to March is less than the evapotranspiration loss, and the resulting soil moisture deficit affects tea bushes. If this dry spell persists for a longer period, tea plants suffer heavily and crop goes down in spite of having sufficient rainfall in the monsoons. Thus, adequate rainfall during winter and early spring is crucial for high yield.

The effect of monsoon dynamics and weather on tea production by using provincial-level data of tea production in China and found yield to be more sensitive to precipitation than to temperature. An increase in the retreat date of the monsoon and an increase in monsoon precipitation are associated with a decrease in tea yield (Mamun 2011a, b; Mamun and Ahmed 2011; Ali et al. 2014; Ahmed and Ahmed 2015; Ali et al. 2015; Rahman et al. 2017). The monsoon season is predicted to be longer with an earlier arrival and a similar or later retreat date. The increase in seasonal mean precipitation is likely to be most pronounced in the study area, while the change in other monsoon regions is subject to greater uncertainty (BCAS 1994; Seran et al. 1999; Bekhit 2006; Yang et al. 2006). Changes in the monsoon season will affect tea production because the quantity and variability of rainfall are crucial. The effects of precipitation relative to temperature are even more important. Diminishing rainfall reduces tea yields, but this depends on its distribution over time. During long rains, tea production is lower when compared with short rains. This is due to long rainy periods reducing sunshine and the photosynthesis of tea leaves. Extreme rainfall events such as floods and droughts will also negatively affect tea yield that a reduction of monthly rainfall by 100 mm could reduce productivity of made tea by 30–80 kg/ha/month (Mondal et al. 2004; Chowdhury and Ward 2004; Muthaiya et al. 2013; Li et al. 2015). The reduction in annual rainy days (even with the same quantity of precipitation) and relative humidity, which are closely correlated, will adversely affect tea production. A further reduction in these parameters will certainly reduce tea growth and production. Furthermore, the rise in temperature will increase soil evaporation and plant transpiration, causing water shortage or seasonal drought in areas with low precipitation. Soil water deficits showed a negative correlation with tea yields (Bore et al. 2013), and higher water availability increased the growth of new leaves (Ahmed et al. 2014). The highest tea leaf production per hectare depends on 4000–4600 mm annual rainfall, according to an analysis of field experiment results with weather data in Bangladesh (Ali et al. 2014). Tea generally exhibits a positive interaction between rainfall and temperature because its production depends on stable temperatures and consistent rainfall patterns (Ochieng et al. 2016). Any significant change in temperature and precipitation will affect production.

Effect of Rainfall on Tea Leaf Production

Rainfall was positively related with tea leaf production in Lackatoorah and Burjan tea estate. In the previous 10 years, the minimum rainfall was recorded 3132.24 mm, whereas the maximum rainfall was recorded 5523.76 mm for Lackatoorah tea estate. However, in the last 10 years, the minimum rainfall was recorded as 3070.22 mm for Burjan tea estate, while the maximum rainfall was recorded as 5050.95 mm. The tea leaf production and rainfall in Loobacherra tea estate were moderately correlated and had positive influence on each other (Mondal et al. 2004; Chowdhury and Ward 2004; Muthaiya et al. 2013; Li et al. 2015). In the past 10 years, the minimum rainfall in Loobacherra tea estate was 3314.54 mm where the maximum rainfall was 5398.77 mm. In this tea estate, the per unit area production was highest among the four tea estate. The soil property and other extraneous factor played positive role for tea leaf production. On the other hand, for good-quality tea, one bud and two leaves are generally collected, but in the field observation, it found that more than two leaves including buds were collected from this tea estate. Malnichera is the oldest tea estate in Bangladesh where there was less fluctuation of tea production per unit area (BCAS 1994; Seran et al. 1999; Bekhit 2006; Yang et al. 2006) (Fig. 6.3).

The figure shows that the maximum rainfall (31.57 mm) was recorded in 2018 though the yield was the lowest (11.24 kg/ha). The production of tea decreases at very high rainfall due to the lack of sunshine (Islam et al. 2005; Ahmmed 2012; Ahmad and Hossain 2013; Hassan 2014; Hossain 2015). Excess water may negatively influence the production of tea due to the saturation of soil, failure of absorption of water by plants. The rainfall was declining from 2012 to 2016. Despite highest rainfall in 2017, the production decreased as 2457.7 mm rainfall was recorded in the summer, while 895.4 mm in the winter. A drastic change in the distribution of month-wise rainfall was observed IPCC (2013) reported that South Asian rainfall is subject to greater uncertainty (BCAS 1994; Seran et al. 1999; Bekhit 2006; Yang et al. 2006). The average rainfall in the month of monsoonal May reduced by 176.6 mm, where increased in the warmest April by 236.8 mm. Tea production is lower in long rains when compared with short rains. This is due to long rainy periods reducing the photosynthesis of the leaves with the decrease sunshine. Extreme rainfall negatively affects tea yield (Wijeratne et al. 2007; Esham and Garforth 2013; Duncan et al. 2016) (Fig. 6.4).

In last 10 years, the minimum rainfall was recorded as 3132.87 mm, whereas the maximum rainfall was recorded as 4692.88 mm. Tea leaf production was slightly increasing with increasing rainfall. The production was fluctuating in the same range of rainfall. It might be potential of the management and other extraneous

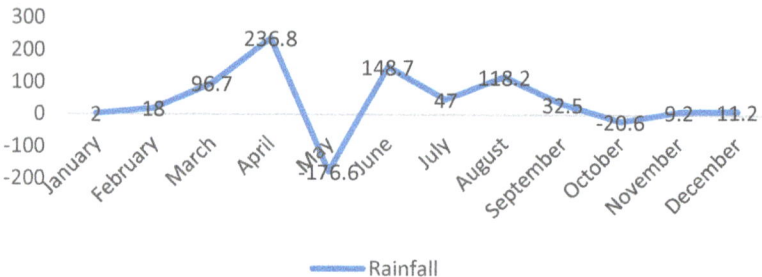

Fig. 6.4 Changes in rainfall month-wise rainfall distribution pattern over 2012–2018

factors like intensity of sunlight, distribution of yearly rainfall, etc. (Mondal et al. 2004; Chowdhury and Ward 2004; Muthaiya et al. 2013; Li et al. 2015). The maximum tea leaf production per hectare of the Lackatoorah tea estate was 791.82 kg which lies in the rainfall 4163.21 mm, and for the Burjan tea estate, it was 1059.91 kg with maintaining rainfall 4666.13. The maximum tea leaf production of other two tea estates, namely Loobacherra and Malnichera was 2605.61 kg/ha and 1563.32 kg/ha with maintaining rainfall 4495.22 mm and 4692.88 mm, respectively. The average rainfall of Lackatoorah, Burjan, Malnichera, and Loobacherra tea estates are 3764.674 mm, 3934.926 mm, 4258.091 mm, 4274.65 mm respectively.

Tea Production Humidification and Humidity Control

A humid climate favors tea growth in the plantation and the right relative humidity level is essential to convert the picked leaves into quality tea during processing (Mamun 2011a, b; Mamun and Ahmed 2011; Ali et al. 2014; Ahmed and Ahmed 2015; Ali et al. 2015; Rahman et al. 2017). During oxidation (also called fermentation), careful control of temperature and humidity is particularly important. This process provides the right conditions for the polyphenols in withered leaves to develop into the aflavins, the arugibins and other compounds that create the characteristic flavors and color of black teas, Pouchong and oolongs (BCAS 1994; Seran et al. 1999; Bekhit 2006; Yang et al. 2006). If leaves dry, oxidation slows as water is required for the process. An ambient relative humidity of 95–98% RH at 20–26 °C is required to maintain fermentation, and even a small drop in humidity level will affect the process, reducing yield. Humidification of other stages, such as in tea bag production to reduce electrostatic problems, also improve productivity.

Adaptation and Mitigation Strategies

Planning for climate change adaptation and mitigation initiatives is essential, not only for dealing with the negative impacts of climate change, but also in order to create cost-effective opportunities and benefits for sustainable development of the tea industry (Mamun 2011a, b; Mamun and Ahmed 2011; Ali et al. 2014; Ahmed and Ahmed 2015; Ali et al. 2015; Rahman et al. 2017).. These strategies should have at least three levels: government policy, technology and technical development, and community involvement for the extension of adaptation and mitigation measures. These should be integrated for the best outcomes (BCAS 1994; Seran et al. 1999; Bekhit 2006; Yang et al. 2006). Climate change is often referred to as a global problem, requiring top-down international and national strategies to achieve substantial climate change adaptation and mitigation. Effectively integrated and coordinated government policies or strategies are required for cost-effectiveness and consistency of implementation.

Strengthening Investment in Field Infrastructure

Infrastructure construction and improvement, such as drainage and irrigation systems, road construction, ecosystem diversity and rebalancing should be strengthened. Infrastructure is considered a very good investment from the cost–benefit analysis point of view, even in the absence of climate change (Islam et al. 2005; Ahmmed 2012; Ahmad and Hossain 2013; Hassan 2014; Hossain 2015). Taking into account the high cost across the board to individual' tea plantations, governments should play a key role in this area.

Good Agricultural Practices (GAP)

GAP and/or green farming consists of a collection of principles which take into account economic, social, and environmental sustainability. These apply to on-farm production and post-production processes for safe and healthy food and non-food agricultural products (Islam et al. 2005; Ahmmed 2012; Ahmad and Hossain 2013; Hassan 2014; Hossain 2015). They also apply to integrated pest management, integrated plant nutrient management, and conservation agriculture, which are beneficial techniques for mitigation and adaption to climate change.

Improving Soil and Water Capacity

Climate change has a negative effect on the basic elements of food production, such as soil, water, and biodiversity. Most tea fields are located on the rain-fed slopes of mountainous areas in which tea yields depend not only on the amount of rainfall, but also on its utilization (Mamun 2011a, b; Mamun and Ahmed 2011; Ali et al. 2014; Ahmed and Ahmed 2015; Ali et al. 2015; Rahman et al. 2017). With increasing temperatures and high evaporation and transpiration, drought will probably be a normal phenomenon in the coming years. Therefore, to develop soil and water conservation measures, increasing soil water-holding capacity will be very important in reducing the impact of drought and maintaining tea production (Mondal et al. 2004; Chowdhury and Ward 2004; Muthaiya et al. 2013; Li et al. 2015). Besides establishing contour terraces, mulching, planting cover crops and installing contour-staggered trenches, ecosystem diversity, and water conservation agents should be considered for further research and development. Adopting small-scale irrigation in tea fields will also increase resilience to drought.

Creating Awareness of Climate Change and Its Impact

Public awareness of climate change and its impact on the tea industry is one of the key strategies in implementing effective participatory climate adaptation on a large scale. Public access to information on climate change and its effects, especially those specific to the region of target audiences, should be made available in local communities.

Conclusion

The study provides evidence that ultimate per hectare tea leaf production is slightly increasing in different tea estates of Sylhet district due to increased rainfall. Heavy or scanty or delayed rainfall adversely affects the growth and yield of tea, but it has been found that tea leaf production is slightly increasing with increase in annual rainfall. Temperature and humidity has no direct effect on tea leaf production in Sylhet region. These study shows that mean annual rainfall and yield per hectare is positively correlated. There are significant uncertainties in the climatic parameters. The present study suggests that if climate changes due to low rainfall and significant increase in temperature will be resulted in a significant loss of yield of tea leaf production.

References

Ahmad I, Hossain MA (2013) Present status of Bangladesh tea sector in respect of world tea: an overview. J Appl Sci Technol 9(1)

Ahmed M (2014) Bangladesh tea-status of current production and research and future development plan. In: Proceedings on international tea symposium 2014, pp 68–80. Hangzhou, China

Ahmed M, Ahmed T (2015) A case study on tea production at northern Bangladesh. Tea J Bangladesh 44:10–18

Ahmmed MF (2012) Promoting equity and access to WaSH among the tea laborers in Sylhet. Institute of Development Affairs (IDEA) and WaterAid, Dhaka

Alam AFM (2002) Research on varietal improvement of tea and their utilization in the tea industry of Bangladesh. In: Proceedings of the international seminar on "varietal development of tea in Bangladesh", 13 July 2002. Organized by Bangladesh Tea Research Institute, Srimangal-3210, Maulvibazar, Bangladesh, pp 7–26.

Ali M, Islam M, Saha N, Kanan AH (2014) Effects of microclimatic parameters on tea leaf production in different tea estates in Bangladesh. World J Agric Sci 10:134–140

Ali M, Uddin M, Mobin M, Saha N (2015) Effects of microclimatic parameter on tea leaf production in different tea estates, Bangladesh. J Environ Sci Nat Resour 7(1):183–188. https://doi.org/10.3329/jesnr.v7i1.22168

Anon A (1996) Climate change scenarios for the Australian region/Climate impact group. CSIRO Division of Atmospheric Research. Australia: CSIRO Division of Atmospheric Research

Banerjee B (1992) Botanical classification of tea. In: Tea cultivation to consumption. In: KC, Clifford MN (eds) pp 25–51. Chapman and Hall, London

BBS (2000) Statistical year book of Bangladesh. Bangladesh Bureau of Statistics. Ministry of Planning, Government of People's Republic of Bangladesh, pp 81–327. Dhaka, Bangladesh

BCAS (1994) Vulnerability of Bangladesh to climate change and sea level rise: concepts and tools for calculating risk in integrated coastal zone management. Technical Report, Bangladesh Center for Advanced Studies, Dhaka 1(2)

Bekhit MY (2006) Levels of essential and non- essential metals in leaves of the tea plant (*Camellia sinensis L.*) and soils of Wushwush farms, Ethiopia. http://etd.aau.edu.et/dspace/bitstream/123456789/307/1/Michael%20Yemane.pdf. Accessed on 10 July 2012\

Boehm R, Cash SB, Anderson BT, Ahmed S, Griffin TS, Robbat A, Stepp JR, Han WY, Hazel M, Orians CM (2016) Association between empirically estimated monsoon dynamics and other weather factors and historical tea yields in China: results from a yield response model. Climate 4:20

Bore JK, BC (2013) Long term impact of climate change on tea yields. Tea 13:57–67

BTB (2017) Monthly bulletin of Statistics on tea, January 2017. Bangladesh Tea Board, Nasirabad, Chittagong

Bangladesh Tea Board (2009) Composition of environmental statistics of Bangladesh. 2005, Bangladesh Bureau of Statistics; Planning Division, Ministry of Planning, Government of the People's Republic of Bangladesh, Dhaka, Bangladesh, p 419

BTD (Bangladesh Tea Board) (2012) http://www.Teaboard.gov.bd/. Accessed 25 June 2012

BTRI (Bangladesh Tea Research Institute) (2012) Brief note on tea culture for the BTRI annual report, Shromongal, Bangladesh

BTRI (2003) Biennial report. Bangladesh Tea Research Institute, Government of People's Republic of Bangladesh, pp 94–98, Srimangal, Moulvibazar, Bangladesh

Butt MS, Ahmad RS, Sultan MT, Qayyum MM, Naz A (2015) Green tea and anticancer perspectives: updates from last decade. Crit Rev Food Sci Nutr 55(6):792–805. https://doi.org/10.1080/10408398.2012.680205

Cai XM, Sun XLL, Dong WXX, Wang GCC, Chen ZMM (2013) Herbivore species, infestation time, and herbivore density affect induced volatiles in tea plants. Chemoecology 24:1–14

Carr MKV, Stephens W (1992) Climate weather and the yield of tea. In: Willson KC, Clifford MN (eds) Tea: cultivation to consumption. Chapmann and Hall, London, pp 87–135

Chen W, Cheung YK (1987). A new approach for the hybrid element method. Int J Numeric Method Eng 1697–1709. https://doi.org/10.1002/NME.1620240907

Cheruiyot EK, Mumera LM, Ng'Etich WK, Hassanali A, Wachira F (2010) High fertilizer rates increase susceptibility of tea to water stress. J Plant Nutrition 33:115–129

Chomchalow N (1996) Herbal tea, an Editorial. NANMAP-17, February 1996, FAO/RAP, Bangkok, Thailand

Chow KB, Kramer I (1990) All the tea in chin. China: China Books & Periodicals

Chowdhury MR, Ward MN (2004) Hydro-meteorological variability in the greater Ganges-Brahmaputra Meghna (GBM) basin. Int J Climatol 24:1495–1508

Duncan JMA, Saikia SD, Gupta N, Biggs EM (2016) Observing climate impacts on tea yield in Assam, India. Appl Geo 64–71. https://doi.org/10.1016/j.apgeog.2016.10.004

Dubravina GA, Zaitseva SM, Zagoskina NV (2005) Changes in formation and localization of phenolic compounds in the tissues of European and Canadian yew during dedifferentiation in vitro. Russ J Plant Physiol 52:672–678

Esham M, Garforth C (2013) Climate change and agricultural adaptation in Sri Lanka: a review. Climate Develop 5(1). https://doi.org/10.1080/17565529.2012.762333

Forrest GI (1969) Studies on the polyphenol metabolism of tissue cultures derived from the tea plant (Camellia sinensis L.). Biochem J 113:765–772

Friedman M (2007) Overview of antibacterial, antitoxin, antiviral, and antifungal activities of tea flavonoids and teas. Mol Nutr Food Res 51:116–134

Gamborg OL, Miller RA, Ojima K (1968) Nutrient requirements of suspension culture of Soybean root cells. Exp Cell Res 50:151–158

Gheyas IA, Sabur SA (2002) Effect of various stimuli on consumer behaviour for food commodities. In: An area of mymensingh distri. Bangladesh J Agricultural Econ,Bangladesh Agricultural Univ 25(2). https://doi.org/10.22004/ag.econ.201453

Gurusubramanian AR (2008) Pesticide usage pattern in tea ecosystem, their retrospects and alternative measures. J Environ Bio 29(6):813–826. https://doi.org/10.1038/s41598-021-82454-3

Haque M (2013) Life in the labour lines: situation of tea workers, environmental governance: emerging challenges for Bangladesh. AHDPH, Dhaka, pp 391–414

Hara Y, SL (1995) Special issue on tea. Food Rev Int 11(3):371–374

Harbowy ME, Balentine DA, Davies AP (1997). Tea Chemistry. Critical Rev Plant Sci 16(5):415–480. https://doi.org/10.1080/07352689709701956

Hassan AE (2014) Deplorable living conditions of female workers: a study in a tea garden of Bangladesh. Am J Human Soc Sci 2(2):121–132

Hays SM, Aylward LL, LaKind JS, Bartels MJ (2008) Guidelines for the derivation of biomonitoring equivalents: report from the biomonitoring equivalents Expert workshop. Regulator Toxicol Pharmacol 51:S4–S15. https://doi.org/10.1016/j.yrtph.2008.05.004

Hicks JRA (2001) RcoA has pleiotropic effects on Aspergillus nidulans cellular development. Mole Microbio 39(6):1482–1493. https://doi.org/10.1046/j.1365-2958.2001.02332.x

Hossain SM (2015) Wage pattern and livelihood of tea garden laborer: a study on Loobacherra Tea Estate, Kanaighat, Sylhet, Bangladesh. Doctoral dissertation, BRAC University

Hossain MA, Uddin S, Hossain MS (2013) Investigate the challenges in tourism business: a study based on sylhet division in Bangladesh. J Sci Technol 11:144–150

IPCC (2013) Climate change (2013) the physical sciences basis, pp 33–118. Cambridge University Press, Cambridge

IPCC (2014) Climate change (2014) mitigation of climate change, contribution of working group III to the fifth assessment report of the intergovernmental panel on climate change. Cambridge University Press, Cambridge, United Kingdom

Islam GMR et al (2005). Present status and future needs of tea industry in Bangladesh. Proc Pakistan Acad Sci 42G (4.M):.3R0.5 I-S3la1m4. 2et0 0a5

Islam GMR, Iqbal M, Quddus KG, Ali MY (2005) Present status and future needs of tea industry in Bangladesh. Proc Pakistan Acad Sci 42(4):305–314

ITC (2015) Annual bulletin of statistics (2015). International Tea Committee (ITC), London, UK, p 158

Jalmi SK, Bhagat PK, Verma D, Noryang S (2018) Traversing the Links between heavy metal stress and plant signaling. Frontiers Plant Sci 9. https://doi.org/10.3389/fpls.2018.00012

Kfoury N, Morimoto J, Kern A, Scott ER, Orians CM, Ahmed S, Griffin T, Cash SB, Stepp JR, Xue D, Long C, Robbat A (2018) Striking changes in tea metabolites due to elevational effects. Food Chem 264:334–341

Khisa P, And Iqbal I (2001) Tea manufacturing in Bangladesh: problems and prospects. In: Proceedings of the international conference on "mechanical engineering", 26–28 December 2001, Department of Mechanical Engineering, Bangladesh University of Science and Technology, Dhaka

Li CF, Zhu Y, Yu Y, Zhao QY, Wang SJ, Wang XC et al (2015) Global transcriptome and gene regulation network for secondary metabolite biosynthesis of tea plant (Camellia sinensis). BMC Genomics 16:560. https://doi.org/10.1186/s12864-015-1773-0

Mamun MSA (2011a) Integrated approaches to tea pest management in South India: a way of sustainable tea cultivation. LAP LAMBERT Academic Publishing GmbH & Co. KG Dudweiler Landstr. 99, 66123 Saarbrücken, Germany, p 59. ISBN: 978–3–8465–5088–5

Mamun MSA (2011b) Development of tea science and tea industry in Bangladesh and advances of plant extracts in tea pest management. Int J Sustain Agril Tech 7(5):40–46

Mamun MSA, Ahmed M (2011) Integrated pest management in tea: prospects and future strategies in Bangladesh. J Plant Prot Sci 3(2):1–13

Mehrin N, Chowdhury T, Nath SR (2016) Operational challenges in providing primary education services in wetland (Haor) and tea garden areas of Sylhet Division in Bangladesh

Mondal TK, Bhattacharya A, Laxmikumaran M, Ahuja PS (2004) Recent advances of tea (Camellia sinensis) biotechnology. Plant Cell, Tissue Organ Cult 76:195–254

Muthaiya MJ, Nagella P, Thiruvengadam M, Mandal AKA (2013) Enhancement of the productivity of tea (Camellia sinensis) secondary metabolites in cell suspension cultures using pathway inducers. J Crop Sci Biotechnol 16(2):143–149. https://doi.org/10.1007/s12892-012-0124-9

Nasir T, Shamsuddhoa M (2011) Tea production, consumption and exports: Bangladesh perspective. Int J Educ Res Technol 2(1):68–73

Ochieng J, Kirimi L, Mathenge M (2016) Effects of climate variability and change on agricultural production: the case of small scale farmers in Kenya. NJAS - Wageningen J Life Sci 71–78. https://doi.org/10.1016/j.njas.2016.03.005

Paul AL, Sng NJ, Zupanska AK, Krishnamurthy A (2017) Genetic dissection of the arabidopsis spaceflight transcriptome: are some responses dispensable for the physiological adaptation of plants to spaceflight? PLoS ONE12(6). https://doi.org/10.1371/journal.pone.0180186.

Prakash O, Sood A, Sharma M, Ahuja PS (1999) Grafting micro propagated tea (Camellia sinensis (L.) O. Kuntze) shoots on tea seedling- a new approach to tea propagation. Plant Cell Rep 18:137–142

Rahman A (2016) An enquiry into the living conditions of tea garden workers of Bangladesh: a case study of Khan Tea Estate. Doctoral dissertation, BRAC University

Rahman MM, Islam MN, Hossain MR, Ali MA (2017) Statistical association between temperature-rainfall and tea yield at Sylhet Malnicherra tea estate: an empirical analysis

Saha (2010) Economics of tea, global tea statistics and tea marketing system in Bangladesh. BTRI, Srimangal-3200, Bangladesh, pp 5–7

Seran TH, Hirimburegama K, Hirimburegama WK, Shanmugarajah V (1999) Callus formation in another culture of tea clones, Camellia sinensis (L.) O. Kuntze. J Nat Sci Found Sri Lanka 27 (3):165–175

Sharma VS, Ramachandran KV, Venkata Ram CS (1981) Tipping in relation to pruning height and its effect on the yield of tea (Camellia spp.). J Plant Crops 9:112–118

Singh SN, Narain A, Kumar P (2006) Socio-economic and political problems of tea garden workers: a study of Assam. Mittal Publications

Singh AK, Bisen JS, Bora DK, Kumar R, Bera B (2012) Comparative study of organic, inorganic and integrated plant nutrient supply on the yield of Darjeeling tea and soil health. Field Crop Res 58:58–61

Smith RH (2012) Callus induction. In: Plant tissue culture: techniques and experiments, 3rd edn. Academic press, pp 63–79. https://doi.org/10.1016/B978-0-12-415920-4.00006-2

Sumi RS (2020) Demand analysis of domestic tea market in Bangladesh: an empirical investigation. Doctoral dissertation, University of Dhaka

Taylor SJ, McDowel IJ (1991) Rapid classification by HPLC of plant pigments in fresh tea (camellia sinensis l) leaf. J Sci Food Agri 57:287–291. https://doi.org/10.1002/jsfa.2740570212

Wijeratne AA, Madawala A (2007) Assessment of impact of climate change on productivity of tea (Camellia sinensis L.) plantations in Sri Lanka. J Nat Sci Found SriLanka 35(2):19–126. https://doi.org/10.4038/jnsfsr.v35i2.3676

Yang CS, Lambert JD, Ju J, Lu G, And Sang S (2006) Tea and cancer prevention: molecular

Zhuang W (1988) A discussion on the history of tea in China. China: Science Press

Chapter 7
Determinants of Food Security in the Environmentally Stressed Areas in Bangladesh

Mohammad Amirul Islam, Avik Chowdhury, Md. Anisuzzaman, Md. Shohanur Rahaman Shetu, Khandaker Md. Mostafizur Rahman, and Moupia Rahman

Abstract The studies identified the strategies that include short-term dietary changes; reducing or rationing consumption; altering household composition; altering intrahousehold distribution of food; depletion of stores; increased use of credit for consumption purposes; increased reliance on wild food; short-term labour migration; short-term alterations in crop and livestock production patterns; pledging, mortgaging and sales of assets; and distress migration. Davies (IDS Bull 24:60–72, 1993) makes the distinction between coping strategies (fallback mechanisms to deal with a short-term insufficiency of food) and adaptive strategies. The communities were selected for the following reasons: firstly, they fall within the coastal saline area of Bangladesh where there are frequent shortages of food due to uncertainty of rainfall and lack of fresh irrigation water. Secondly, the area provides an opportunity to study impacts associated with climate change and vulnerability on crop and livestock. This analysis revealed that the households not having connection with several NGOs or GOs were 0.028 times significantly less likely to be food secure than the households having that contact. The analysis depicted that the households not having the access to marketing information were 0.214 times significantly less like to be food secure as compared to the households having the same

M. A. Islam (✉) · K. Md.M. Rahman
Department of Agricultural Statistics, Bangladesh Agricultural University, Mymensingh, Bangladesh

A. Chowdhury
Bangladesh Bank, Dhaka, Bangladesh

Md.Anisuzzaman
Stuttgart, Germany

Md.S. R. Shetu
PKSF, Dhaka, Bangladesh

M. Rahman
Department of Environmental Science and Management, North South University, Dhaka, Bangladesh

© Springer Nature Switzerland AG 2021
Md. Jakariya and Md. N. Islam (eds.), *Climate Change in Bangladesh*,
Springer Climate, https://doi.org/10.1007/978-3-030-75825-7_7

access to the information related to market price of input, output, materials needed for shrimp culture. Provision of training, in this study, was significantly associated with the food security status. Regression analysis had shown that households who had not received training related to shrimp culture management were 0.370 times significantly less likely to be food secure than the households who had received that training.

Keywords Climate change · Food security · Distress migration · Vulnerability · Bangladesh

Introduction

Food security has been an important development agenda since the inception of the Millennium Development Goals (MDGs) and got refocused in the Sustainable Development Goals (SDGs). Government of Bangladesh has been taking the issue of food security along with poverty very seriously. Since then, analysis of food security is getting more and more attention among the researchers. A number of studies (for example, Imam et al. 2018; Bala et al. 2012, 2013; Dash 2005; Halder and Mosley 2004; Hossain 1989; Kundu 2004; Radhakrishna and Ravi 2003; Rahman and Khan 2005; Rahman et al. 2005; Amin and Farid 2005; Talukder and Quilkey 1991) focusing on different aspects of food insecurity have been conducted that identified lack of economic and social access to safe and nutritious food items to meet daily dietary needs as the major reason for food insecurity. The studies find out that lack of employment in a particular locality can create a situation of food insecurity through trimming down income, the key to economic access to food. And if this situation prevails for a longer time, the society will inevitably face famine, hunger and malnutrition. The overall situation of food security in Bangladesh has always been fragile because of mostly the constant and steady gap between supply (availability) and demand (need) of major food grains due to many reasons like inadequate production, improper distribution, lack of food aid and importing capacity (MWCA 2006). These studies did not deal much with the coping strategies of the vulnerable people suffering from lack of food security but were related to similar aspects of insecurity such as lack of physical, social and economic access, food intake and lack of employment. Rahman et al. (2009) in their study examined the impact of paid employment and self-employment on income and prospects of household food security. The results of the multiple regression on household income show that among the three status of employment, those who were in self-employment had the highest income. The dummy variables for wage employment and salaried employment had negative coefficients (self-employment is the base group). Further, the findings of the logit regressions show that wage employment raises food insecurity (defined as a combination of below poverty line income and calorie adequacy). The situation is much worse among wage workers in rural areas. In addition, the study explores that the production of rice within the household makes a positive impact on calorie consumption as the proportion of households that consumes less than 2122 kcal is lower for the rice-producing households than for their non-producing counterparts. Similar relationship holds for non-poor households. Food security is historically attributed to the overall regional, national, or even global food supply and shortfalls in supply compared to requirements, but, with increased observation of disparities in the sufficiency of food intake by certain groups, despite overall adequacy of supply, the term has been applied recently mostly at a local, household or individual level (Foster 1992) and has been extended beyond notions of food supply to include elements of access (Sen 1981), vulnerability (Watts and Bohle 1993) and sustainability (Chambers 1989). Most definitions of food security vary around that proposed by the World Bank (1986); major components of the most common definitions are summed up by Maxwell and Frankenberger (1992) as 'secure access at all times to sufficient food for a healthy life'. Summarizing the conceptual literature on food security, Maxwell and Frankenberger conclude four steps. First, 'enough' food is mostly defined with emphasis on calories, and on requirements for an active, healthy life rather than simple survival—although this assessment may at the end be subjective. Second, access to food is determined by food entitlements (Sen 1981), which are derived from human and physical capital, assets and stores, access to common property resources and a variety of social contracts at household, community and state levels. Third, the risk of entitlement failure determines the level of vulnerability and hence the level of food insecurity, with risk being greater, the higher the share of resources devoted to food acquisition. And finally, food insecurity can exist on a permanent basis (chronic) or on a temporary basis (transitory) or in cycles. A full definition of food security thus includes the related concepts of access, sufficiency, security (or vulnerability) and sustainability.

The 1996 World Food Summit emphasized on three important components in relation to ensuring food security, e.g. availability of adequate food, stability in food supplies, access to food and nutrition security (FAO 1996). The poor, especially the ultra-poor, suffer from food insecurity basically because of lack of purchasing capacity and fewer opportunities to have easy access to available food. The landless people living in the Third World countries are the main victims of food insecurity because of multiple reasons. These landless people are mostly forced to become day labourers who are dependent on casually occasional earning for their livelihood. Millions of such people are affected and suffer from chronic and transitory food insecurity due to seasonal variation in agricultural activities and limited opportunities to have employment in non-agricultural activities. Household-level food insecurity is very common among these people. When there is no earning, there is no economic access to food at home and this unfortunate situation compels these food-insecure people to survive on taking cash money as loan from the moneylenders at a high interest rate (Amin and Farid 2005). In many parts of Bangladesh, the people living in low-lying areas face frequent natural disasters and lead an uncertain life because of chronic food insecurity. Household-level food insecurity is accepted as a part and parcel of everyday life in these areas, and the people are often forced to depend on moneylenders or food lenders for very survival

(Amin and Farid 2005). Needless to say that it is poverty, which generally manifests the situation of food insecurity for the people. Around 44% of the total population in Bangladesh lives in poverty, measured by CBN method (BBS 2007). Bangladesh is one of the most overpopulated countries in the world, and it remains heavily dependent on agrarian economy (58.3% people directly dependent on agriculture); a mismatch of resources in terms of inappropriate use and distribution leaves more than 60 million people living below poverty line with a very high level of child malnutrition (Rajaretnam and Hallad 2000); though absolute figures and percentages have changed positively underlying reasons are still the same. Another reason for food insecurity and poverty could be described as landlessness of a considerable number of people. Landlessness and lack of employment opportunities in the rural areas force the rural people to migrate to the urban areas causing a chaotic situation in the big cities, especially the capital city, Dhaka. The number of landless people is increasing gradually and has become tripled in the last five decades. At present, more than half of the people are landless, while it was only 28% in 1972 (BBS 2004). Of course, Bangladesh did not witness famine for a long time after 1974, but starvation is very often found among the chronically poor people who specially suffer during lean season (before harvesting of *Aman* paddy). The poor people consume less amount of food, suffer from under nutrition to hunger, do not have access to basic health services, and suffer from illiteracy and many other deprivations. Insufficient food intake causes malnutrition which is very much prevalent in the households or areas hit by periodic or frequent food shortages. Mothers and children often suffer because of male biased food distribution in the household. Women of childbearing age and their under-five children become the ultimate victims of hunger and malnutrition (Rajaretnam and Hallad 2000). Girls and women, for various reasons, are discriminated and heavily malnourished in comparison with boys and men (PRSP 2005). Around 50% of the children are underweight, indicating the severity of malnutrition prevalent in the country. Bangladesh government and World Food Programme (WFP) have jointly been implementing Vulnerable Group Development (VGD) programme to ensure food security and to enhance the development of the poorest 5% of the rural women since 1980. The women having VGD card are supposed to receive 15 kg rice or wheat per head as monthly ration. These are some of the strategies provided to the food deficient people to survive. The local government is often given the responsibility to identify and select the deserving women for VGD (UNDP 2003).

Bangladesh has continued to demonstrate a steady increase in the domestic production of food grain since 1971, and the production has increased from 11 million metric tons in the 1970s to 24.7 million metric tons in 2003–2004 (WFP 2005). However, the net domestic production is not sufficient to meet the requirement of cereals by the increasing number of population in the country. Every year the remaining food gap is met by import of food grain, which is gradually increasing. It has been reported that during 2004–2005, the shortage of food grains in the country was nearly 1.9 million tons (BBS 2004). So, the production of food grains could never meet the original demand of a huge population living in a small country like Bangladesh. Food insecurity has in fact been reduced in Bangladesh compared to the situation prevailed in 1970s. But it is far from being over. Although people do not die of hunger these days, a considerable portion of the population always remains hungry because of lack of food security. More than 60 million people are still found suffering from chronic hunger in Bangladesh which is definitely larger than those in many other countries in the world. Bangladesh has, therefore, the third largest poor population across the globe after China and India (UNDP 2005). There are more or less 27 food security and social safety net programmes existing in the country, but they are insufficient to cover the increasing number of the hungry people (PRSP 2005). These programmes generally do not include transitory poor, and the effectiveness of these programmes has often been questioned because of leakage and misappropriation (World Bank 2004). These facts are reflected in the malnutrition situation, especially for those living in rural areas, as their nutritional status shows. The Bangladesh Bureau of Statistics (BBS) in its Household Income and Expenditure Survey 2005 shows that the malnutrition problem for the poorest is quite severe as 14% of the rural people consume less than 1600 cal per capita per day, necessary for bare survival; another 10% consume between 1600 to 1800 cal per day, and around 23% consume more than 1800 cal, but certainly less than the minimum requirement of 2122 cal per capita per day considered to be standard in Bangladesh. Each calorie group is termed as 'ultra-food deficit', 'hard-core food deficit' and 'moderately food deficit', respectively (BBS 2003). Other indicators have been used to monitor food security, including food balance sheets, rainfall and marketing data, and even anthropometric measurement (Maxwell and Frankenberger 1992). Haddad et al. (1994) documented a variety of indirect indicators that can be used as predictors for food insecurity at the household level, including asset ownership, household size and dependency ratio. While they mostly discuss the use of single indicators, they suggest that indicators could be combined for greater specificity. Several authors analysed the use of and reliance upon, strategies for dealing with insufficiency of food at the household level as direct indicators (Watts 1983; de Garine and Harrison 1988; Corbett 1988; Reardon et al. 1988; De Waal 1989; Drèze and Sen 1989; Moris 1989; Frankenberger and Goldstein 1990; Leonard 1991; Rahmato 1991; Teklu 1992; de Garine 1993; Davies 1993; Frankenberger and Coyle 1993; Devereux and Naeraa 1993; Watts and Bohle 1993; Eele 1994). The studies identified the strategies that include short-term dietary changes; reducing or rationing consumption; altering household composition; altering intrahousehold distribution of food; depletion of stores; increased use of credit for consumption purposes; increased reliance on wild food; short-term labour migration; short-term alterations in crop and livestock production patterns; pledging, mortgaging and sales of assets; and distress migration. Davies (1993) makes the distinction between coping strategies (fallback mechanisms to deal with a short-term insufficiency of food) and adaptive strategies (long-term or permanent changes in the way in which households and individuals acquire sufficient food or income).

Bangladesh has been characterized by many agro-ecological zones. Also, the country, being a deltaic one, has been facing serious environmental conditions in different parts. The major areas where there are prevailing environmentally stressed conditions are Chittagong Hill tracts in the south-east, Haor areas in the north-east, coastal areas in the south-west, drought-affected areas in the north-west and river erosion-prone areas across the river basins. Ample interventions have been made in the drought-prone areas (Zug 2006; Hossain and Haque 2010), and studies suggested that the so-called Monga that was prevailing in the areas during dry seasons is now absent. Bala et al. (2012, 2013) attempted to explore the food security condition and strategies for the development in the Chittagong Hill tracts. In the contrary, number of studies conducted to understand the food security condition in the remaining three vulnerable areas are scarce. This research, hence, put focus on the remaining three environmentally stressed areas, namely Haor area, coastal saline-prone areas and areas suffering from river erosion. In fact, this article is a compilation of three independent researches covering the three areas.

Data

Data for Haor Areas

Four randomly selected districts were selected, namely Netrakona, Habiganj, Sunamganj and Kishoreganj. These four districts had 28 upazilas containing wetland. Among these there were several upazilas which had partly *haor* areas, partly non-*haor* areas. Besides these, there were some upazilas which were fully attributed with the characteristics of *haor* area. From these types of upazilas, ten (10) upazilas were randomly obtained from four districts, which were the best representatives of *haor* area. The ultimately selected ten upazilas were Khaliajury, Modon and Mohonganj from Netrakona, Ajmiriganj from Habiganj, Derai, Sulla, Jamalganj, and Dharmapasha from Sunamganj, Itna and Mithamain from Kishoreganj. Total sample size from this area was 150.

Data for Coastal Areas

The data for coastal area was collected from the four selected upazillas of Khulna in Bangladesh. They were: Batiaghata, Dumuria, Dacope and Paikgacha. These upazillas were selected purposively to achieve the objectives of the study. These study areas were selected to represent the communities that are affected by climate uncertainty, tropical cyclones and storm surges to different degrees. The communities were selected for the following reasons: firstly, they fall within the coastal saline area of Bangladesh where there are frequent shortages of food due to uncertainty of rainfall and lack of fresh irrigation water. Secondly, the area provides an opportunity to study impacts associated with climate change and vulnerability on crop and livestock. The sample size from this area was 120.

Data for River Erosion Areas

There are many rivers in Bangladesh, the river-eroded areas are almost available in our country. However, Sirajgonj district is one of the most important areas characterized by the proneness of river erosion and the destruction of current years. The data for river erosion areas were collected from four upazillas of Sirajgonj district in Bangladesh. They were Sirajgonj Sadar, Kazipur, Belkuchi and Shahjadpur. The sample size for this area was 120.

Direct Calorie Intake (DCI) Method

The direct calorie intake method estimates the per capita calorie intake at household level. In this method, the food consumed during the last seven days in a household was first averaged and then the average content of food per day per household was converted into kilocalorie. Afterwards, the amount of calorie intake was converted into per capita per day. According to this method, a household is considered as 'food insecure' with less than 2122 kcal per day (Talukder and Quilkey 1991).

Single-Level Binary Logistic Regression Model

Let Y be a dichotomous dependent variable, say change in poverty situation taking values 0 and 1 and suppose that $Y = 1$, if there is a positive change and $Y = 0$, otherwise. Also let X be an independent variable. Then the form of the logistic regression model is:

$$P = p(Y = 1|X) = \frac{1}{1 + e^{-(\beta_0 + \beta_1 X)}}, \quad \text{and} \quad 1 - P = p(Y = 0|X) = \frac{1}{1 + e^{\beta_0 + \beta_1 X}}$$

Then, a transformation of P known as the logit transformation is defined as

$$g(x) = \text{logit } P = \log\left[\frac{P}{1 - P}\right] = \beta_0 + \beta_1 X$$

There are many desirable properties of this transformation $g(x)$. The logit $g(x)$ is linear in its parameters. It may be continuous and may range from $-\infty$ to $+\infty$.

Depending on the range of x for more than one independent variable, the model can be generalized as:

$$g(x) = \text{logit}(P_i) = \beta_0 + \sum_{l=1}^{k} \beta_l X_{il}, \quad \text{where,} \quad l = 1, 2, \ldots, k \text{ and } i = 1, 2, \ldots, n$$

Dependent Variables

Food security status of the farmers was considered as the dependent variable for this study. Scores are assigned as 1 and 0 to the responses of 'food secure' and 'food insecure', respectively. Note that the food security status was calculated by using appropriate method based on the information provided by the individual respondent.

Results

Our survey data reveals that about 37.3% of the households were food insecure in Haor areas. This was followed by 28.3% in coastal areas and 41.6% in the river erosion areas (Fig. 7.1). Note that the percentages may be different than that of a national level study because of the difference in sampling techniques and sample sizes.

Determinants of Food Security in Hoar Areas

Four independent variables among eight have been turned out to be statistically significant (Table 7.1). Significant variables are: source of capital used in crop cultivation, profession of the household head, farm size and the level of off-farm income. Participation in safety net programmes, education, family size and age of household head have, however, been appeared to be insignificant in the model. Source of capital used in the agricultural production had been found to be significant (at 5% level of significance) as a factor that might have powerful impact on determining the food security status of the household members. Operating the rice cultivation with borrowed capital instead of own capital was supposed to throw the

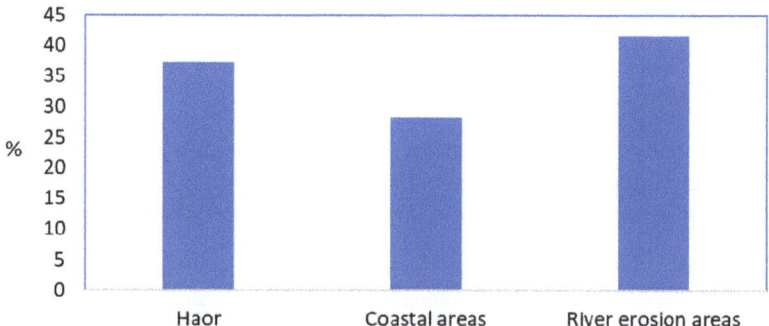

Fig. 7.1 Percent of food-insecure people in the study regions

family out of the food security. In such of cases, family members were 0.32 time less likely to achieve food security than the families used its own household capital to cultivate rice. This result was, obviously, expected. Borrowing capital means increasing risk. The increased amount of risk compelled the families to be vulnerable in terms of food security. Very often, a borrowing family was not able to afford all the necessary inputs timely, and this way they were also at risk of under-production which, in turn, might lead to the food insecurity.

The variable of primary occupation of the household head has also been come out to be a significant determinant of food security (at 5% level of significance). Shifting into business or fishing from agriculture was expected to enhance the opportunity to ensure food security by almost 51%. This result pointed out the current vulnerability of paddy growers across the country. Nowadays, throughout the country, we observed the despair and depression in the eyes of most rice farmers. They claimed that they could not cope with the increased price of input materials. Often, rice output could not meet the cost of production. Thus, shifting to

Table 7.1 Binary logistic regression model for determination of factors influencing food security in Haor areas

Independent variables	Coefficient	S.E.	Odds ratio
Intercept	12.959	9.725	2.382
Participation in safety net program (r: No)			
Yes	−0.349	0.501	0.706
Source of capital (r: own household)			
Borrowed	−1.139*	0.475	0.320
Both	−0.635	1.028	0.138
Age of household head (r: 15–30 years)			
31–50 years	0.578	0.469	1.782
51 years and above	−0.317	1.252	0.728
Education of household head (r: up to class three)			
Class four or more	−0.685	0.586	0.504
Profession (r: agriculture)			
Business or fishing	0.596*	0.240	1.51
Government employee or others	0.449	6.49	1.566
Family size (r: 1–4)			
5 and above	−0.654	9.205	0.520
Farm size (r: marginal)			
Small and medium	0.113	0.731	1.119
Large	1.965*	0.759	7.132
Off-farm income (r: no off-farm income)			
≤ Tk. 20,000	0.101	1.354	1.107
>Tk. 20,000	0.704*	0.260	2.023

Note r denotes reference category; *$p < 0.05$

business or fishing could be a powerful weapon to battle with food insecurity in the *haor* area.

The result showed that the size of the farm was a significant factor (at 5% level of significance) of elevating a family to food secure status. The families of large farms with more than 749 decimal of land were 7.13 times more likely to become food secure than that of a marginal farm with less than 50 decimal of land. The result was also expected. More land implied the increased amount of wealth which was an indicator of affluence. The affluent families were, reasonably, safe in terms of food security.

Off-farm income of the families had been found to be another influencing factor of food security status determination. Families generating more than Tk. 20,000 as off-farm income were almost two times more likely to obtain food security than the families earned nothing as off-farm income. Off-farm income supplied additional amount of money to the households. So, the families could increase its food expenditure and could meet up all the necessary dietary needs. Consequently, these families achieved food security.

Determinants of Food Security in Coastal Areas

Household's access to food depends on household income, asset, remittance, gifts, borrowing, income transfer and also sometimes food aid. Increased incomes of household can improve household food security in terms of improved access to food. In addition, expanded asset reduces the vulnerability of households to short-term disruption in income flows and help to prevent degradation of household food security in times of adversity (NFP 2006).

Among the variables considered in regression analysis, seven variables, in particular, amount of land holdings, number of family member, access to credit, contact with GO/NGO, access to marketing information, and receiving training had significant effect on the food security status. Regression result had shown that farmers who had been shifted to shrimp culture from rice production were 3.92 times significantly more likely to be food secure than the rice producers (Table 7.2).

The regression analysis showed that the households having land holdings of above 750 decimal were 9.07 times significantly more likely to be food secure compared to the households having land of 5 decimal to 49 decimal (reference category). The binary logistic regression had shown that the households not having access to credit were 0.252 times significantly less likely to achieve the status of food security than the households having the access to credit from various sources. Contact with the NGO or GO was significantly related to households' food security status. This analysis revealed that the households not having connection with several NGOs or GOs were 0.028 times significantly less likely to be food secure than the households having that contact.

The analysis depicted that the households not having the access to marketing information were 0.214 times significantly less like to be food secure as compared

Table 7.2 Binary logistic regression model for the determination of factors influencing food security status in coastal areas

Independents variables	Coefficient	S.E	Odds ratio
Intercept	5.506**	0.771	
Farming types (r: rice producers)			
Shrimp farmers	1.418*	0.687	3.92
Sole shrimp farmers	0.60	0.502	1.32
Amount of land holdings (r: 5–49 decimal)			
50–749 decimal	1.435	0.881	2.81
Above 750 decimal	2.47**	0.761	9.07
Number of family member (r: 1–6)			
7–14	1.21	0.787	1.484
Access to credit (r: yes)			
No	−1.44*	0.614	0.252
Contact with GO/NGO (r: yes)			
No	−2.3**	0.793	0.028
Access to marketing information related to shrimp farming (r: yes)			
No	−1.657*	0.821	0.214
Receiving of training (r: yes)			
No	−2.687**	0.911	0.370

Note r denotes reference category; $*p < 0.05$ and $**p < 0.01$ are the levels of significance

to the households having the same access to the information related to market price of input, output, materials needed for shrimp culture. Provision of training in this study was significantly associated with the food security status. Regression analysis had shown that households who had not received training related to shrimp culture management were 0.370 times significantly less likely to be food secure than the households who had received that training.

Determinants of Food Security in River Erosion Areas

The time duration from shifting was an important factor that influenced food security status of the displaced people. From Table 7.3, it is seen that the likelihood to be food secure for those who shifted for the years ranging 6–10 were 4.46 times compared to the households with shifting years less than 6. As the duration of the time after shifting increased, the affected people got enough time to be stable. They started to do something for their living which was consistent and could start their lives with newly arranged way; hence, the more time they got the more secure they became.

Table 7.3 Binary logistic regression model for determining the factors affecting food security in river erosion areas

Independent variables	Coefficient	S.E	Odds ratio
Intercept	1.411	0.903	4.099
Cultivable land (value) (r: up to Tk. 50,000)			
Tk. 50,001–Tk. 100,000	0.311	1.020	1.365
Above Tk. 100,000	1.251	1.405	3.947
Duration after shifting (r: less than 5 years)			
Between 6–10 years	1.496**	0.513	4.462
Above 10 years	1.544	1.842	4.681
Number of times of shifting (r: single time)			
More than once	−1.118	4.019	0.327
Age of the household head (r: below 46 years)			
Above 46 years	−0.968	0.580	0.380
Education of the household head (r: up to primary)			
Above primary	0.341	0.940	1.407
Family size (r: up to 5)			
Above 5	−1.907*	0.812	0.149
Losses due to erosion (r: less than Tk. 75,000)			
Between Tk. 75,001–150,000	−4.629	2.460	0.010
Above Tk. 150,000	-2.728*	1.06	0.093
Annual income (r: less than Tk. 90,000)			
Between Tk. 90,001–140,000	2.22**	0.644	9.218
Above Tk. 140,000	2.839	1.28	9.389
Annual expenditure (r: less than Tk. 90,000)			
More than Tk. 90,000	1.347	1.065	3.847
Profession (r: agriculture)			
Private job	1.722*	0.741	2.059
Others (business or integrated)	1.108**	0.481	3.029

Note r denotes reference category; $*p < 0.05$ and $**p < 0.01$ are the levels of significance

The effect of family size was found to be significant and negative (at 5% level). Families with more than five members were almost 85% less likely to be food secure than families with less than five members. The demand for the food of the large family is always higher than the demand of the smaller one. With the number of members increased, in most of the cases, the number earning person remained the same (as a result of lack of enough employment opportunity), which meant, with increasing family size they became more vulnerable economically.

Income was always a vital factor in any socio-economic research. In this study, as a determinant of food security, income was found highly significant (at 1% level). Income between Tk. 90,000 to 140,000 could make a household nine times food secure compared to the household with income less than Tk. 90,000 per year.

To buy more food, there was no alternative to income, hence with the increased income, food security level increased, which was predictable.

As mentioned earlier, the main profession of people of these areas was agriculture. Profession is another important factor that influenced the food security status. When the profession of the household's head changed to private job (mainly as garments worker), it significantly (at 5% level) makes households two times food secure compared to the households where the profession of the household's head was agriculture based. The security level became three times when the profession changed to something other than agriculture and private job (including business or some other integrated professions). The awareness worked behind this scenario. A farmer living in a village was obviously less aware than a person worked in a garment factory in a town. Also, income of a person engaged in agriculture is normally less than the person having a profession other than agriculture. After being displaced, people also lost their agricultural land in huge amount that compelled them to lead a life with lower standard.

Conclusion

This article reveals the food security condition in three major environmentally stressed areas, namely Haor areas, coastal areas and river erosion areas. The determinants of food security were also identified, which have been summarized in this section. A significant portion of total households of Haor areas were experienced food insecurity. With a view to mitigating the severe consequences of malnutrition, proper and prompt attempts should be taken by the government. Shifting to business or fishing from agriculture was supposed to help ensure the food security. Operating large farm (more than 750 decimals) was also useful to improve food security status. Increase of off-farm income was also expected to be helpful to make people food secure.

The possible consequences of climate change, i.e. sea-level rise, salinity intrusion and unpredictable weather have substantial influences on present cropping pattern, farming practices and eventually on the livelihood of the people in coastal area. For instance, rice production has been reduced due to the inability of the rice varieties to cope with unfavourable climatic conditions. But, rice production has paramount importance on the people and on the overall economy of the country. For averting potential losses, farmers in a large number have been shifting to shrimp culture. It was evident from the overall findings of the study that the shifting had significant and positive impact on food security, moreover on overall betterment of the people living in disaster-prone coastal area. The study had provided some valuable indicators of food security status in coastal zone of Khulna district. A large portion of households were experiencing calorie-based food insecurity.

Poverty and food insecurity are both the causes and consequences for each other. In river erosion areas, everyone was cursed with poverty. There were many factors reasonable for the food insecurity in these areas. Lack of education, large family size, and low level of income were the main factors to lead them to insecurity. With food insecurity, their social status and health status became miserable.

It is apparent that three environmentally stressed areas are facing different types of problems resulting differential impact on the food security condition of the households living in these areas. Hence, policies should be devised separately for different areas keeping the set of determinants identified. A common food security programme will not produce good outcome. Tailor-made area-specific programmes will perform better to eradicate food insecurity from these areas, and subsequently, it will impact on the achievement of related SDG goals.

References

Amin MR, Farid N (2005) Food security and access to food: present status and future perspective. Paper presented at the national workshop on food security in Bangladesh, organized by GoB and WFP

Bala BK, Hossain MA, Majumder S (2012) Food security and ecological footprint of Chittagong hill tracts in Bangladesh. Science Vision 18(1&2)

Bala BK, Majumder S, Altaf Hossain SM, Haque MA, Hossain MA (2013) Exploring development strategies of agricultural systems of Hill Tracts of Chittagong in Bangladesh. Environ Dev Sustain 15(4):949–966

BBS (2003) Household income and expenditure survey-2000. Bangladesh Bureau of statistics, Ministry of Planning, GoB, Dhaka

BBS (2004) Report of the poverty monitoring survey 2004. Bangladesh Bureau of Statistics, Ministry of Planning, GoB, Dhaka

BBS (2007) Household income and expenditure survey-2005. Bangladesh Bureau of Statistics, Ministry of Planning, GoB, Dhaka

Chambers R (1989) Vulnerability, coping, and policy. IDS Bull 20(2):1–7

Corbett JEM (1988) Famine and household coping strategies. World Dev 16(9):1099–1112

Dash BP (2005) Regional food security experience: lessons learnt from India and Timor Leste. Paper presented at the national workshop on food security in Bangladesh, organized by GoB and WFP

Davies S (1993) Are coping strategies a cop out? IDS Bull 24(4):60–72

de Garine I (1993) Coping strategies in case of hunger of the most vulnerable groups among the Massa and Mussey of northern Cameroon. GeoJournal 30(2):159–166

de Garine I, Harrison GA (1988) Coping with uncertainty in food supply. Clarendon Press, Oxford

de Waal A (1989) Famine that kills. Clarendon Press, Oxford

Devereux S, Naeraa T (1993) Drought and entitlement decline in Namibian Agriculture, 1992. Social Science Division, University of Namibia, Windhoek, Namibia, Mimeo

Drèze J, Sen A (1989) Hunger and public action. Clarendon Press, Oxford

Eele G (1994) Indicators for food security and nutrition monitoring: a review of experience from southern Africa. Food Policy 19(3):314–328

FAO (1996) World Food Summit 1996. Rome, organized by Food and Agriculture Organization of the United Nations, November 1996

Foster P (1992) The world food problem: tackling the causes of undernutrition in the Third World. Lynne Rienner Publishers, Boulder

Frankenberger T, Goldstein DM (1990) Food security, coping strategies and environmental degradation. Arid Lands Newsletter 30:21–27

Frankenberger T, Coyle PE (1993) Integrating household food security into farming systems research/extension. J Farming Syst Res Extension 4(1):35–65

Haddad L, Kennedy E, Sullivan J (1994) Choice of indicators for food security and nutrition monitoring. Food Policy 19(3):329–343

Halder SR, Mosley P (2004) Working with the ultra-poor: learning from Brac experiences. J Int Dev 16:387–406

Hossain M (1989) Food security, agriculture and the economy: the next 25 years. University Press Limited, Dhaka

Hossain MA, Haque ME (2010) Food security and income generation through access to common property waterbodies in Monga-affected areas of Bangladesh. Published by Department of Aquaculture, Bangabandhu Sheikh Mujibur Rahman Agricultural University, Gazipur

Imam MF, Islam MA, Hossain MJ (2018) Factors affecting poverty in rural Bangladesh: An analysis using multilevel modelling. J Bangladesh Agr Univ 16(1):123–130

Kundu A (2004) Food security systems in India: analysis of conceptual issues in contemporary policy debate. Paper presented in workshop on social safety nets India jointly by WFP and World Bank in September, New Delhi

Leonard W (1991) Household-level strategies for protecting children from seasonal food scarcity. Soc Sci Med 33(10):1127–1134

Maxwell S, Frankenberger T (1992) Household food security: concepts, indicators, measurements; a technical review. International Fund for Agricultural Development/United Nations Children's Fund, Rome

Moris J (1989) Indigenous versus introduced solutions to food stress in Africa in seasonal variation in third world agriculture: the consequences for food Security. Johns Hopkins University Press for the International Food Policy Research Institute

MWCA (2006) Impact Survey (Cycle 2003–2004) of FSVGD project. Department of Women's Affairs, Ministry of Women and Children Affairs, World Food Programme, In Partnership with European Commission (Report written by Dr. M.Z. Hossain)

NFP (2006) National food policy, Ministry of food and disaster management, Government of Bangladesh

PRSP (2005) BANGLADESH—unlocking the potential. National Strategy for Accelerated Poverty Reduction, General Economics Division, Planning Commission, Government of Bangladesh

Radhakrishna R, Ravi SC (2003) Malnutrition in India: trends, and determinants. Economic and Political Weekly, December 29.

Rahman MM, Khan SI (2005) Food security in Bangladesh: food availability. Paper presented at the national workshop on food security in Bangladesh, organized by GoB and WFP

Rahman RI, Begum A, Bhuyan HR (2009) Impact of paid employment and self-employment on income and prospects of household food security. A research conducted by Bangladesh Institute of Development Studies under NFPCSP of GoB and FAO, Dhaka

Rahman SM, Hoque A, Talukder RA (2005) Food security in Bangladesh: utilization, nutrition and food safety. Paper presented at the national workshop on food security in Bangladesh, organized by GoB and WFP

Rahmato, D. (1991) Famine and survival: a case study from northeast Ethiopia. Uppsala

Rajaretnam T, Hallad JS (2000) Determinants of nutritional status of young children in India: an analysis of 1992–93 NFHS data. Demography-India 29(2):BBS (2004)

Reardon T, Matlon P, Delgado C (1988) Coping with household-level food insecurity in drought-affected areas of Burkina Faso. World Dev 16(9):1065–1074

Sen AK (1981) Poverty and famines: an essay on entitlement and deprivation. Clarendon Press, Oxford

Talukder RK, Quilkey JJ (1991) Food preference and calorie intake behaviour in Bangladesh. Bangladesh J Agric Econ 16:1–26

Teklu T (1992) Household response to declining food entitlement: The experience in western Sudan. Q J Int Agric 31(3):247–261

UNDP (2003) Human development report 2003. Oxford University Press, New York

UNDP (2005) Human Development report 2005. Oxford University Press, New York

Watts, M. (1983) Silent violence: food, famine, and peasantry in northern Nigeria. University of California Press, Berkeley, California

Watts M, Bohle H (1993) Hunger, famine, and the space of vulnerability. GeoJournal 30(2):117–126

WFP (2005) Food availability and consumption situation in the country. Bangladesh food security brief, World Food Program

World Bank (1986) Poverty and hunger: issues and options for food security in developing countries. A World Bank Policy Study, Washington, D.C.

World Bank (2004) World development indicators 2004. World Bank

Zug S (2006) Monga—seasonal food insecurity in Bangladesh—bringing the information together. The Journal of Social Studies, No. 111, July-Sept. 2006, Centre for Social Studies, Dhaka

Chapter 8
Climate Change and Municipal Solid Waste Management in Dhaka Megacity in Bangladesh

Hassan Mahmud

Abstract Urban growth resulted from economic development led to the multiplication of the per capita solid waste generation rate and volume/quantity of solid waste. Expansion of metropolitan area poses a greater risk toward the environment which needs immediate attention to the problem of solid waste disposal, air pollution control and deterioration of the urban environment. The average highest generation rate was found to be 0.368 kg/capita/day at residential areas in Dhaka, whereas the lowest was 0.259 kg/capita/day in Barisal. The mean generation rate in residential areas was obtained as 0.309 kg/capita/day for six cities of different income level. However, they estimated that overall waste generation rate based on population was ranging from 0.325 to 0.485 g/cap/day, while highest rate is 0.485 kg/cap/day in Dhaka city. Scarcity of land in Bangladesh makes this option unaffordable. The land value of Matual and Amin Bazar landfills presently used by DSCC and DNCC, respectively, is very high. Considering minimum land price of Tk. 50 million per acre and maximum 10 m height of waste piles, only land value per ton of waste disposal would be Tk. 2471/ton of waste considering density of 500 kg MSW/m^3. The biogas would be used to generate electric power or used as biogas for other purposes. The project will also eliminate breeding ground for the pests and reduce health hazard, air and water pollution and threat to human health and the environment.

Keywords Climate change · Megacity · Waste management · Urban environment · Bangladesh

H. Mahmud (✉)
Department of Environmental Science and Management, North South University, Dhaka, Bangladesh
e-mail: hassan.mahmud@northsouth.edu

© Springer Nature Switzerland AG 2021
Md. Jakariya and Md. N. Islam (eds.), *Climate Change in Bangladesh*,
Springer Climate, https://doi.org/10.1007/978-3-030-75825-7_8

Introduction

With 1265 people per square kilometre, Bangladesh is one of the most densely populated countries in the world (Worldometers 2017). It has a staggering population of 162.7 million people, according to the latest publication on November 8, 2017, and is increasing at a rate of 1.05% annually based on the latest United Nations estimates. The GDP of Bangladesh has been growing at a high average annual rate of above 6% since 2006. Due to this stable economic growth, it is included in the "Next Eleven" emerging countries which are expected to follow BRICs in the economic development. As a result of rapid population growth and urbanization, a humongous part of the population became the urban population. For a better understanding of the subject, Tables 8.1 and 8.2, respectively, provide historical statistics and an insight into the national population and urban population of the country.

Urban growth resulted from economic development led to the multiplication of the per capita solid waste generation rate and volume/quantity of solid waste. Expansion of metropolitan area poses a greater risk toward the environment which needs immediate attention to the problem of solid waste disposal, air pollution control and deterioration of the urban environment. Dhaka, the capital of Bangladesh, is ranked ninth among the megacities of the world (UNESCAP 2009), compassing 55,500 people per sq. km (UN-HABITAT 2013) and ranking fourth densest city in the world (list of cities by population density, Wikipedia). Forbes (2013) estimated that in 2013, the population of Dhaka was nearly 14.4 million within a city area of 165.63 km^2.

As shown in Table 8.1, the urban population in Bangladesh is increased rapidly. The rapid population growth, particularly in the cities of Dhaka North and Dhaka South, which were created in 2011 through the division of the capital of Bangladesh, and in Chittagong, has turned solid waste management into a significant social problem, as these cities are producing increasing amounts of solid waste. Dhaka North City Corporation (DNCC) and Dhaka South City Corporation (DSCC) are responsible for solid waste management in Dhaka North and Dhaka South, respectively.

Cities can be dense without being overpopulated. But Dhaka is a dysfunctional megacity which is manifested by an inefficient solid waste management system. A substantial proportion of solid waste is disposed of in open spaces, causing severe environmental hazards and public discomfort. A study by the Department of Environment (DoE) found thhat Dhaka's solid waste management is negligent in comparison with other developing countries (DoE 2013). According to Global Air Report 2017, Dhaka is the second most polluted city in the world with the alarming inefficient control municipal solid waste.

Table 8.1 Population of Bangladesh (2018 and historical)

Year	Population	Yearly % change	Yearly change	Density (P/km²)	Urban pop %	Urban population	Country's share of world pop	World population	Bangladesh global rank
2018	166,368,149	1.03	1,698,398	1278	35.30	60,649,009	2.18%	7,632,819,325	8
2017	164,669,751	1.05	1,718,191	1265	35.70	58,746,319	2.18%	7,550,262,101	8
2016	162,951,560	1.09	1,750,674	1252	34.90	56,856,665	2.18%	7,466,964,280	8
2015	161,200,886	1.16	1,810,357	1238	34.10	54,983,919	2.18%	7,383,008,820	8
2010	152,149,102	1.19	1,743,600	1169	30.30	46,035,276	2.19%	6,958,169,159	8
2005	143,431,101	1.74	2,369,972	1102	26.80	38,373,642	2.19%	6,542,159,383	8
2000	131,581,243	2.08	2,574,874	1011	23.70	31,229,852	2.14%	6,145,006,989	8
1995	118,706,871	2.25	2,503,646	912	21.90	26,003,685	2.06%	5,751,474,416	9
1990	106,188,642	2.64	2,597,755	816	20.00	21,274,633	1.99%	5,330,943,460	9
1985	93,199,865	2.73	2,345,801	716	17.70	16,496,299	1.91%	4,873,781,796	8
1980	81,470,860	2.70	2,032,987	626	15.00	12,251,656	1.83%	4,458,411,534	8
1975	71,305,923	1.85	1,251,631	548	10.00	7,107,810	1.75%	4,079,087,198	9
1970	65,047,770	3.10	1,842,746	500	7.70	5,034,728	1.76%	3,700,577,650	9
1965	55,834,038	2.98	1,526,858	429	6.40	3,552,536	1.67%	3,339,592,688	12
1960	48,199,747	2.73	1,215,645	370	5.30	2,543,661	1.59%	3,033,212,527	11
1955	42,121,524	2.14	845,369	324	4.80	2,020,726	1.52%	2,772,242,535	

Source Worldometers (2017)

Table 8.2 Bangladesh population forecast

Year	Population	Yearly % change	Yearly change	Density (P/Km²)	Urban pop %	Urban population	Country's share of world pop	World population	Bangladesh global rank
2020	169,775,309	1.04	1,714,885	1304	38.00	64,479,585	2.18	7,795,482,309	8
2025	178,262,909	0.98	1,697,520	1369	41.50	74,020,473	2.18	8,185,613,757	8
2030	185,584,811	0.81	1,464,380	1426	44.80	83,160,000	2.17	8,551,198,644	8
2035	191,600,525	0.64	1,203,143	1472	47.80	91,510,533	2.15	8,892,701,940	8
2040	196,294,312	0.49	938,757	1508	50.40	98,935,284	2.13	9,210,337,004	8
2045	199,743,520	0.35	689,842	1534	53.00	105,950,886	2.10	9,504,209,572	8
2050	201,926,816	0.22	436,659	1551	55.70	112,443,436	2.07	9,771,822,753	8

Source Worldometers (2017)

MSW Generation Rates

BUET (2000) has reported from multiple studies data that MSW generation in DCC from 1985 to 1999 was in the range of 1040 and 5000 ton/day (BKH 1985–86; DCC 1985; LBI 1990; Bhide 1990; MMI 1991; PCI 1991; PAS 1997; RSWC 1998; BCAS 1998; Bhuiyan 1999). According to Bhuiyan (1999), estimation of daily waste generation in the Dhaka City Corporation (DCC) is 3500 tons, out of which 1800 tons are collected and dumped at landfills, 400 tons are piled up on roadsides or open spaces, 400 tons are recycled, and the rest is illegally dumped on the way to dumpsite.

Waste Concern Study (2005) assessed that total waste generated in Dhaka was about 4634.52 tons/day accumulating to over 1.69 million tons/year. Based on the total estimated urban population of Bangladesh in the year 2005, per capita waste generation rate was computed as 0.41 kg/capita/day, and the total waste generated in the urban areas of Bangladesh per day was 13,332.89 tons as shown in Table 8.3.

However, in a study, MSW generation rate was determined as 0.34 kg/day in Dhaka in the year 2004 and also revealed that during the wet season, the rate of waste generation increases by 46% (JICA 2005; Ayatollah et al. 2005). Alamgir and Ahsan (2007a) reported that about 78% solid waste is coming from residential sector and 20% from commercial sector, 1% from the institutional sector and rest from other sectors. Alamgir and Ahsan (2007a) have also estimated a total of 7690 tons of MSW generated daily in the six major cities (Dhaka, Chittagong, Khulna, Rajshahi, Sylhet and Barisal), while the Dhaka city contributed the most (69%, 5340 t) to the total waste stream in 2005. The average highest generation rate was found to be 0.368 kg/capita/day at residential areas in Dhaka, whereas the lowest was 0.259 kg/capita/day in Barisal. The mean generation rate in residential areas was obtained as 0.309 kg/capita/day for six cities of different income level. However, they estimated that overall waste generation rate based on population was ranging from 0.325 to 0.485 g/cap/day, while highest rate is 0.485 kg/cap/day in Dhaka city. This per capita generation rate is lower than that reported by Waste Concern Report of 2005. Chowdhury et al. (2014) assessed Dhaka North City Corporation (DNCC), and Dhaka South City Corporation (DSCC) collectively generates about 1.6 million tons of municipal waste per year.

Ahsan et al. (2014) used the same waste generation rate in various cities of Bangladesh as Waste Concern Report of 2005 ranging from 0.25 to 0.56 kg/cap/day, while the highest waste generation rate is 0.56 kg/cap/day in Dhaka city, and the lowest waste generation rate is 0.25 kg/cap/day in Barisal. The trends of waste generation per year are growing at a rate of 0.1343 million tonnes per year (368 t per day). About 78% of solid waste is generated from housing sector and 20% from business sector, 1% from the institutional sector and the rest from other sectors (Ahsan et al. 2014). The waste generation in municipal areas of Bangladesh anticipated to an increment of 0.6 kg/cap/day by 2025 (Bhuiyan 2010). The waste generation rate has increased from 1.1 million tons in 1970 to 5.2 million tonnes in 2015. However, for 2011–12, much higher annual MSW generation rate of 22.5

Table 8.3 Total waste generated in urban areas of Bangladesh in 2005

City/town	WGR[a] (kg/cap/ day)	No. of city/ town	Total population (2005)	Population[b] (2005)	TWG[c] (Ton/day) Dry season	TWG[c] (Ton/day) Wet season	Average TWG (Ton/day)
Dhaka	0.56	1	61,16,731	67,28,404	3767.91	5501.14	4634.52
Chittagong	0.48	1	23,83,725	26,22,098	1258.61	1837.57	1548.09
Rajshahi	0.3	1	4,25,798	4,68,378	140.51	205.15	172.83
Khulna	0.27	1	8,79,422	9,67,365	261.19	381.34	321.26
Barisal	0.25	1	3,97,281	4,37,009	109.25	159.49	134.38
Sylhet	0.3	1	3,51,724	3,86,896	116.07	169.44	142.76
Pourashavas	0.25	298	13,831,187	1,52,14,306	3803.58	5553.22	4678.42
Other Urban Centers	0.15	218	83,79,647	92,17,612	1382.64	2018.66	1700.65
Total	–	522	3,27,65,516	3,60,42,067	10,839.75	15,826.04	13,332.89

[a]WGR = Waste generation rate
[b]Including 10% increase for floating population
[c]TWG = total waste generation which increases 46% in wet season from dry season
Source 1 JICA (2004), 2 Chittagong City Corporation, 3 Field Survey, 4 Sinha (2000), 5 Field Survey, 6 Sylhet City Corporation, 7, 8 Field Survey
Average per capita urban waste generation rate is estimated as 0.41 kg/capita/day
Source Waste Concern (2005)

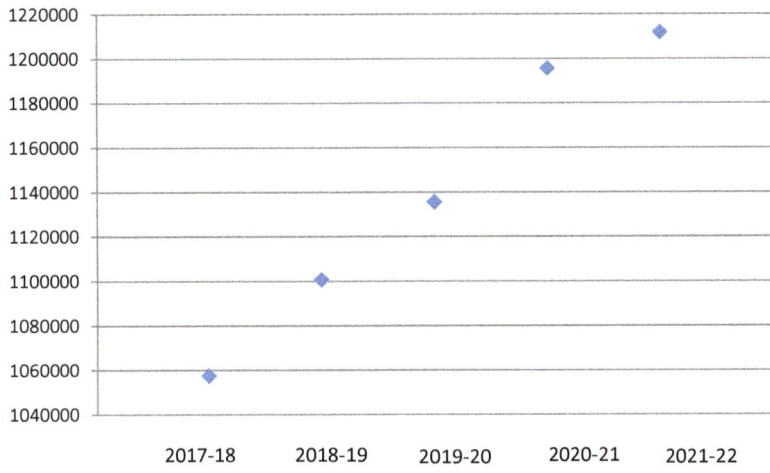

Fig. 8.1 Projected waste generation in tons for DNCC for 2017–18 to 2021–22. *Source* DNCC Waste Report 2016–2017

million tons has been reported by Hossen et al. (2017) from Waste Atlas (2014), which seems very high. High moisture content and low calorific value characterized the solid waste properties in Dhaka city as is high with organic and food waste as observed by Yousuf and Rahman (2007).

DNCC in its Waste Report 2016–2017 reported that DNCC does not know actual waste generation rate but it knows per capita per day waste collection. In 2015–16 and 2016–17, waste collection was 0.391 kg/per day and 0.513 kg/day, respectively. However, the generation rate could be 0.693 kg/per day, calculated by adding 20% uncollected waste and more than 15% average recyclables separated prior reaching the waste at collection facilities not included in collection figure in the DNCC reported collection rate of 0.513 kg/day for 2016–17. Population of DNCC is reported as 4,718,002 in 2016–17. Waste collection was 602,975 tons in year 2014–15, 683,288 tons in year 2015–16 and 852,391 tons in year 2016–17, which shows collection growth rate of 13.30% and 24.77% in 2015–16 and 2016–17, respectively. DNCC also projected waste collection volume for next 5 years from 2017–18 to 2021–22, which is shown in Fig. 8.1.

DNCC in its report 2016–2017 also provided some data on medical waste, electronic waste, construction waste and different types of industrial waste generated in the city. It predicted that e-waste from mobile phones which contains nickel cadmium batteries would be 1033.56 tons in 2017, 1068.26 tons in 2018, 1104.26 tons in 2019, 1141.62 tons in 2020 and 1169.98 tons in 2021.

DSCC also does not know actual waste generation rate, but it provided the total waste collected and disposed based on the recorded weight of waste by weighing scales at Matuail Sanitary Landfill. Based on the recorded measurement at Matuail Sanitary Landfill, DSCC has collected and disposed 643,985 tons in 2015 and 730,946 tons in 2016, which shows collection growth rate of 13.50% over 2015

(Harun 2017). This waste collection growth rate is very close to that of DNCC. Population of DSCC is reported as about 3,100,000 in 2011 and has no proper data after that; however, based on a rough estimate, it is stated that the present population is about 10,000,000 (Islam 2017). Considering about 8,000,000 population in 2016, waste collection rate was 0.25 kg/day in 2016. Main reason for this low collection rate of DSCC compared to DNCC is possibly due to establishment of secondary transfer stations (STS) by DNCC, which has contributed in higher collection rates by DNCC compared to DSCC. DSCC's collection rate possible has not improved more that 50% as was reported for old Dhaka City Corporation by Waste Concern (2005). DSCC could not provide any data on medical waste, electronic waste, construction waste and different types of industrial waste generated in the city. It can be assumed that the generation rate of these would be almost same a DNCC. However, due to more population in DSCC, quantity would be higher.

From the above data, it can be concluded that reported per capita per day generation rates vary widely as the estimates of solid waste generation rate in Bangladesh cities are on the assumption of approximate city populations, do not take into account amount of waste recycled/sold to Feriwallas (vendors) at the sources, recyclables separated and collected by informal sector waste pickers before waste reaches the waste disposal or transfer stations. Also, a substantial portion of the wastes are not collected and are not accounted for during estimation. Moreover, due to change of buying practice such as processed food items such as chicken, fish, and establishment of super shops, which markets processed meat or fish and other semi or fully prepared foods, resulting a substantial quantity of food processing waste not to enter the waste steam as they are sold at the sources and are of high demand. For calculation purpose, it is decided to use generation rate of 0.41 kg/day for the year 2005 (Waste Concern 2005, 2009) and DNCC reported collection rate of 0.513 kg/day for 2016–17, which is 35% less than 0.693 kg/day for 2016–17 for DNCC, calculated as possible generation rate per person per day as shown above under DNCC discussion. This 35% less generation rate would compensate for lower generation rates of other urban areas of Bangladesh as reported in many reports. Thus, yearly increase of per capita per day generation rate would be 1.85% year over year from 2005 to 2017. Based on the above assumptions using the urban population provided in Tables 8.1 and 8.2, yearly MSW generation is shown in Table 8.4 and Fig. 8.2.

Waste Composition

Waste Concern (2009) reported that physical composition of urban solid waste in Bangladesh is 67.65% food waste, 9.73% paper, 5.1% plastics, leather and rubber, 0.26% metal, 1.3% glass and ceramic, 4.5% wood/grass/leaves, 2.5% rags, textile, jute, 0.64% medicines/chemicals and 8.79% rocks, dirt and other waste as shown in Fig. 8.3.

Table 8.4 Waste generation from 2005 to 2030

Year	Urban population	MSW generation, tons/year
2030	83,160,000	19,679,635
2025	74,020,473	15,982,667
2020	64,479,585	12,703,239
2018	60,649,009	11,518,445
2017	58,746,319	10,954,430
2016	56,856,665	10,409,490
2015	54,983,919	9,883,773
2010	46,035,276	7,550,447
2005	38,373,642	5,742,616

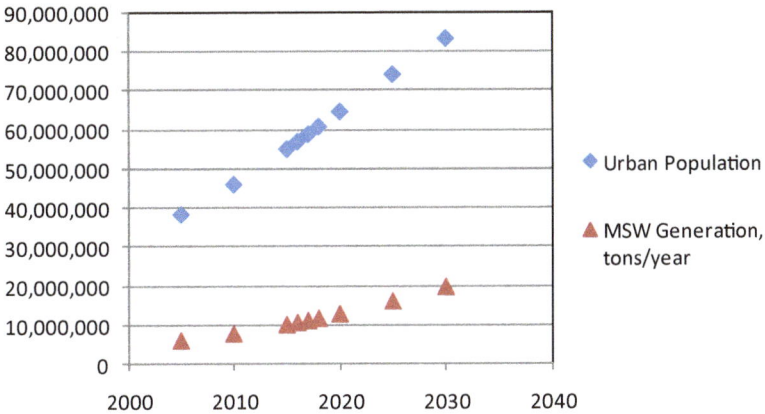

Fig. 8.2 MSW generation from 2005 to 2030

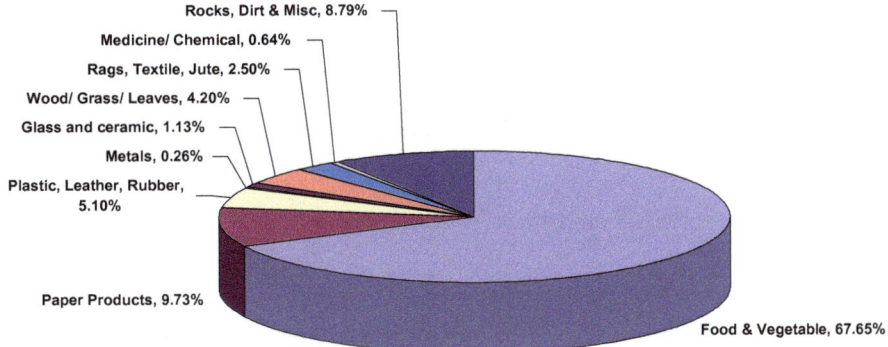

Fig. 8.3 Bangladesh urban waste composition

Islam and Jashimuddin (2017) reported the MSW composition of Chittagong city from six studies (same MSW character of Chittagong city is reported in these studies Alamgir and Ahsan 2007a, b; Uddin and Mojumder 2011; Chowdhury et al. 2013; Salam et al. 2012; Sujauddin et al. 2008) as 61–74% food/organic waste, 3–10% paper, 2–3% plastics, 4–8% textile and wood, 1% leather and rubber, 2% metal, 1–5% glass and 2–11% other waste.

Shams et al. (2017) used the compositions studies of Menikpura et al. (2013), AQUA (2012) and Paul et al. (2014) to obtain an average composition of MSW generated in the six major cities (Dhaka, Chittagong, Khulna, Rajshahi, Sylhet, and Barisal) as 74% food waste, 8% paper, 5% plastics, 2% textile and wood, 2% leather and rubber, 2% metal, 1% glass and 6% other waste.

Rahman et al. (2013) claimed that the waste in Dhaka comprises mainly 60% of organic/food waste.

Summarizing the above reported data on MSW composition of different urban areas of Bangladesh, the composition of the waste has been divided into three major groups based on its characteristics and possible end use and is shown in Table 8.5.

However, the generation rate and composition of residential MSW change with the economic conditions of a country (Tchobanoglous et al. 1993). Bangladesh GDP is increasing fast. As a result, the per capita GDP has increased to $1677 in June 2018, which has lifted Bangladesh to a middle income country. This will reflect in the change in composition of residential MSW, which would decrease the percentage of food waste and increase the percentage recyclables as well as generation rate. This possible composition has been considered in life cycle analysis.

Organic/Biodegradable Waste Diversion or Nutrient and Energy Recovery

From the MSW composition data given above, it is evident that the Bangladesh urban solid waste contains more that 60% food/organic waste. Chowdhury et al. (2014) observed that only a small portion of Dhaka's organic waste is diverted before being transported to the city's landfills or deposited in open spaces. However, one small-scale, private composting company is currently (2014)

Table 8.5 Present average composition of MSW of Bangladesh

Item	Range (%)	Typical (%)
Biodegradable such as food, organic, grass, leaves, non-synthetic textile, rags, jute and wood	60–76	70
Recyclables such as paper, plastics, glass and metals	10–25	15
Others such as rocks and dirt	4–20	15

collecting (fora fee) and processing approximately 100 metric tons of organic waste per day and selling the compost to local farmers (Chowdhury et al. 2014). Dhaka has preliminary plans to develop a larger Integrated Resource Recovery Center at one of the city's landfills to process waste into compost and to generate electricity using landfill gas as reported by Chowdhury et al. (2014). However, only 2% of the total waste is composted and recycled (Waste Concern 2011). The total biodegradable waste can be aerobically composted to produce compost or anaerobically digested to produce biogas and digestate to be used as organic fertilizer/soil conditioner in agriculture.

Recycling

Bhuiyan (1999) reported that 400 tons out of 3500 tons, which is 11.43% of generated MSW in Dhaka CC, was recycled by the informal sector.

The detailed survey by JICA (2005) based on site-specific observations and data collection from the recycle market shows that significant amounts of packaging waste and other valuable materials (plastic, glass, metal, etc.) from MSW are recovered by the scavengers at the kerb side and at the dump site. There are approximately 500 shops engaged in plastic item recycling. The survey found that recycling activity by the informal sector rises to a total of 436 tons/day, which is 8–9% of the total generated.

Waste Concern (2005) reported that informal sector is responsible for recycling from 4 to 15% of the total solid waste generated in different cities and urban centers.

Bari et al. (2012a) reported that in Rajshahi city, Bangladesh, only 8.25% of the total generated waste was recycled in 2010–11, which represented 54.6% of total recyclable wastes and 68.29% of readily recyclable wastes. It was also reported that there were 34 shops involved in buying and selling recyclable waste in Rajshahi City. Bari et al. (2012b) reported that in Khulna city, Bangladesh, only 7.65% of the total generated waste was recycled through 310 shops involved in buying and selling recyclable waste.

As Dhaka city (DNCC and DSCC) generates more than 30% of the total waste of all urban areas of Bangladesh and had a recycling rate of about 15% in 2005 (Waste Concern 2005) and other cities recycle in the range of 4% to 8.25% before 2011. As Bangladesh has achieved per capita GDP of $1602 in 2017, which has lifted Bangladesh to middle income country, it would be safe to consider that an average of 15% waste is recycled in 2017. Establishment of source separation could improve the percent of recycling as presently a large amount of recyclable losses their value due to missing with food waste, which makes it difficult to recover.

Healthcare Waste

Bangladesh Health System Review of 2015 has reported that during 2007–2013, the number of both hospitals and total number of beds in the public sector has steadily increased. The number of beds in PHC facilities at upazila level and below reached 18 880 across 472 facilities in 2013 compared to 16781 across 430 facilities in 2007 and 27,053 in 126 facilities at secondary and tertiary level. In the private sector, there were 2983 registered hospital and clinics, with 45,485 beds (MOHFW 2013). Thus, the total number of functional beds (public and private) in the country is around 91,000 in 2013.

DoE (2010) in a report stated that hospital and clinic is another sector having serious implications for hazardous waste that is growing fast. Comprised of 590 government hospitals, 1809 private hospitals and clinics and 1745 diagnostic laboratories, it has been growing fast, e.g., just between 2007 and 2008 the hospital and clinic facilities' number doubled. Although their locations are spread all over the country, Dhaka accounts for the highest proportion of large facilities with huge number of patients, beds and waste generated. Categorized as general, infectious and sharp, the generation rates of these wastes from hospitals are 1.87, 0.15 and 0.09 kg/bed/day from hospitals. Corresponding proportions for diagnostic laboratories are: 0.032, 0.0033 and 0.0155 kg/bed/day. Combining the infectious and sharp waste, hazardous waste generation rate from these two types of facilities is 0.24 and 0.0188, respectively. As such, total hazardous waste generation per bed per day stands at 0.2588 kg. As per Basel classification, this group contains Y1 (clinical wastes from medical care in hospitals, medical centers and clinics), Y3 (waste pharmaceuticals, drugs and medicines) and H6.2 (infectious substances) types of hazardous waste. Hospitals and clinics have been growing fast that can be sensed from the fact that their number by our estimate stood at 4118 in 2008, which is double than that of 2007 source data. Yearly generation of wastes from this group stood at 12,271 metric tons in 2010. The DoE (2010) report provided corresponding estimate for 2013 as 19,578 metric tons. Taking into account the number of bed of 91,000 in 2013 reported in Bangladesh Health System Review of 2015 using MOHFW data, the medical waste would be 8596.04 tons for 2013, which is only 43.919% of that estimate in the DoE (2010) report.

Table 8.6 Healthcare establishments and medical waste collection in DNCC

Year	HCE	Medical waste collection, tons
2012–13	230	780
2013–14	260	1073
2014–15	328	1218
2015–16	394	1319
2016–17	400	1346

Source Prism Bangladesh Foundation reported in DNCC Waste Report 2016–2017

Prism Bangladesh Foundation (PBF), an NGO, is engaged to collect, treat and dispose Healthcare Waste for DNCC. According to DNCC waste report 2016–17, medical waste producing institute and collection of medical waste is increasing in the following rate shown in Table 8.6.

Only a part of the healthcare waste is collected properly. Not all the cities have any facility to collect or properly treat or dispose such waste. Data shown in the table above provide a clear picture of this statement as medical waste collected in 2016–17 represents only 0.158% of waste collected by DNCC.

E-Waste

E-waste is defined as all types of electrical and electronic equipment (EEE), and its portions that have been discarded by its holder as waste without the intent of reuse. In the study of Global E-waste Monitor, it has been classified as discarded electrical and electronic equipment divided into the following six categories:

(a) Temperature exchange equipment (cooling and freezing equipment),
(b) Screen monitors (televisions, monitors),
(c) Lamps (fluorescent, LED lamps),
(d) Large equipment (washing machines, dishwashing machines),
(e) Small equipment (vacuum cleaners, microwaves),
(f) Small IT and telecommunication equipment (mobile phones).

E-waste contains various materials including hazardous metals like heavy metals (mercury, lead) and chemicals (chlorofluorocarbon, flame retardants). E-waste also contains valuable materials (copper, iron) and precious metals such as gold and silver (Bladé et al. 2015).

As reported by DNCC in its waste report 2016 and 2017, in DNCC, mobile phone rejection was 7.1 million pieces, and thee-waste created from mobile phones was 1033.56 tons in 2017 considering average 0.1456 kg/phone and predicted that it will increase by 3.34% annually to reach to 1169.98 tons in 2021. However, it was reported by Rahman (2017) citing Waste Concern that the rejection in Dhaka city was 2.88 million pieces in 2009 and would increase by 20% annually, which differs from that reported by DNCC. It was also reported that in 2014 mobile phone rejection in Bangladesh was 18 million pieces, which would had generated 1800 tons in 2014 considering 0.1 kg/phone (Rahman 2017). The study also reported that in 2014 computer rejection would be 0.30 million pieces creating a total e-waste of about 4095 tons considering 50% desktop weighing 25 kg/piece and 50% laptop weighing 2.15 kg/piece. Rahman (2017) also reported that 1.20 million of TVs would be rejected which would had generated 36,000 tons of e-waste in 2014 considering 30 kg/TV. It was also stated that no infrastructure has been developed to separate e-waste from other non-hazardous waste by any city corporation including DNCC and DSCC.

Solid Waste Management Perspective and Challenges in Bangladesh

(a) Current and future waste treatment in Bangladesh is greatly affected not only by population growth and consumption patterns, but also by (i) the policies, regulations and incentives provided by governments; (ii) right affordable technology, (iii) applicable funding from private and public sectors and (iv) community social awareness.

(b) At present, a major portion of waste remains untreated in Bangladesh which is about 88%, and about only 50–60% is disposed in sanitary landfills. Only 2% of the total waste is composted and recycled. Today untreated municipal solid waste is mainly disposed in the open landfill sites (Waste Concern 2011).

(c) Out of these total generated wastes, the larger part is organic which can be converted into compost aerobically or biogas and digestate by anaerobic digestion. Today, many cities and towns are operating small-, medium- and large-scale aerobic composting plants in Bangladesh mostly using cow dung and poultry litters. Biogas can be utilized to produce electricity or used for cooking. These technologies should be developed for low cost, simple and easy to run. Industrial scale biogas plants using anaerobic digestion process can be established in the country to generate biogas for cooking and power generation using large quantity of organic fraction of municipal solid waste (OFMSW) co-digested with cow dung, poultry litter, human excreta, slaughter house waste, leachate from waste, agricultural waste, food processing waste and bagasse from sugarcane industries.

(d) Recyclables (plastic, metal, glass, paper, etc.) are mainly recycled informally by small- and medium-sized industries in an unhealthy manner. Local, inefficient polluting technologies are being used by these industries. Few large formal industries are also found to be using recyclables as their raw materials. These SMEs are suffering from lack of modern affordable technological knowhow, training and incentives from the government and proper infrastructures.

(e) With command and control only, the concept 3R (reduce, reuse, recycle) is difficult to implement. At present, most of the polluting industries do not treat their waste and effluent due to several reasons such as lack of incentives, fiscal and financial incentives by the government, lack of awareness and lack of appropriate technological know, etc.

(f) However, the National 3R Strategy for the country gives special attention to selection of appropriate and affordable technology. It says "development and transfer of environmentally sound technologies for waste management and the 3Rs that are applicable in the context of prevailing socio-economic and climatic condition of the country through collaboration among stakeholders such as national governments, local governments, private sectors (including inter-industrial collaboration), consumers, manufactures, informal sectors and research bodies should be promoted. Waste management and treatment intensive industries may be given special attention in this context. Collaboration with materials industry is a key for technical capacity development for the 3R-related technologies and industries. The markets for recycled materials should be stimulated through measures such as standards, incentives, green procurement etc."

(g) Technology from outside should be carefully assessed and verified before bringing it inside the country to avoid problems.

(h) There is need to address the issue of extended producers responsibility (EPR) to minimize problem.

Options of MSWM for Bangladesh

The most used three types of technologies for final disposal/treatment globally after satisfying 3Rs are (a) thermal processes; (b) engineered sanitary land filling and (c) biological degradation processes. Based on the composition shown in Table 8.5, Bangladesh MSW is high in food and organics waste, thus would be high in moisture and would be low in caloric value. This would make Bangladesh MSW not suitable for any thermal process, such as incineration, pyrolysis, gasification or plasma arc processes, should be avoided and would be highly energy negative. Design and operation of an engineered sanitary landfill are expensive and require a large land area. Even with extreme precautions, odor problem cannot be eliminated in full. Scarcity of land in Bangladesh makes this option unaffordable. The land value of Matual and Amin Bazar landfills presently used by DSCC and DNCC, respectively, is very high. Considering minimum land price of Tk. 50 million per acre and maximum 10 m height of waste piles, only land value per ton of waste disposal would be Tk. 2471/ton of waste considering density of 500 kg MSW/m^3. This value does not incorporate cost of engineered sanitary landfill construction and operation, which is very expensive. On the other hand, high organic content makes Bangladesh MSW highly suitable for biological degradation processes and could be best treatment and disposal option. Within the biological processes, aerobic composting would be net energy negative due to oxygen requirement and would need 50 to 70 kW/ton composted, whereas anaerobic digestion would be net energy positive and would provide about 200 kW/ton digested. Organic fertilizer (digestate) obtained from anaerobic digestion is much better in quality than the organic fertilizer (compost) obtained from aerobic composting due to the loss of nitrogen during aeration. Leachate and odor are other problems of aerobic composting. On the other hand, due to the closed nature of the anaerobic digestion process, no nitrogen or any other nutrients are lost rather concentrate in digestate, and the slurry generated from this process is considered a nutrient-rich organic fertilizer. Therefore, anaerobic digestion could be the best available technology for management of MSW generated in the cities of Bangladesh. Although the initial capital cost would be higher than that of aerobic composting, the payback period would be much faster with or without CDM benefit and has a low operating cost. Due to operational simplicity and energy consumption, dry high solid batch anaerobic digestion is better than wet low solid anaerobic digestion. Therefore, dry high solid batch anaerobic digestion is recommended for Bangladesh, which will produce solid and liquid organic fertilizer and biogas. The biogas would be used to generate electric power or used as biogas for other purposes. The project will also eliminate breeding ground for the pests and reduce health hazard, air and water pollution and threat to human health and the environment.

Life Cycle Analysis of MSWM in Bangladesh

Life cycle analysis based on all the data and analysis performed on MSWM in Bangladesh is shown in Fig. 8.4.

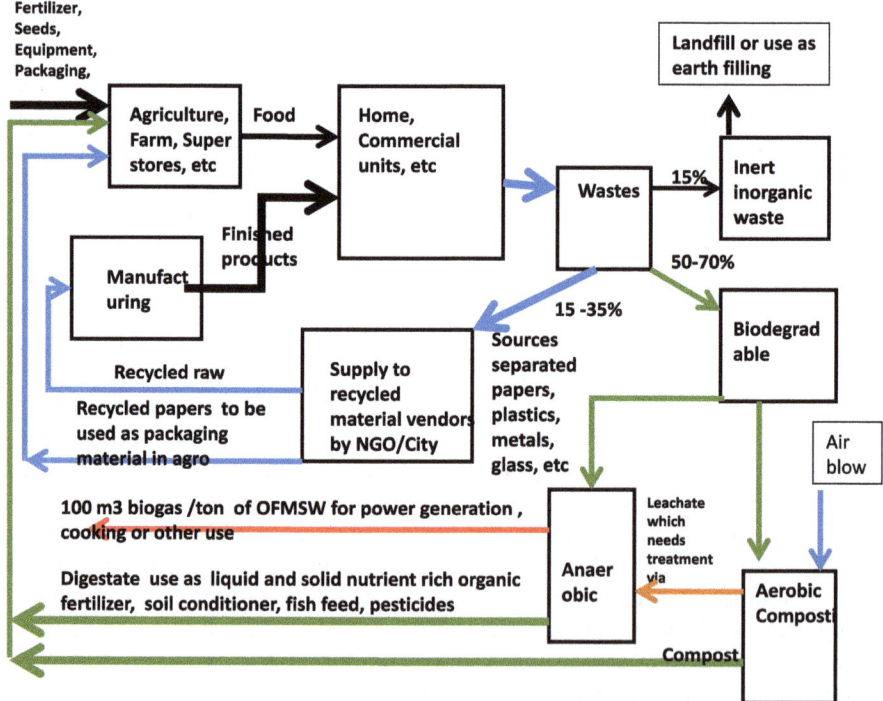

Fig. 8.4 Projected life cycle analysis for MSW of Bangladesh

Green House Gas Reduction Through Biodegradation Processes

Aerobic composting and anaerobic digestion are acknowledged by the UNFCCC as emission reduction methodologies related to organic waste (methodology number AM0025, www.unfccc.int/methodologies/PAmethodologies/index.html). The method depends on the differences between the global warming potential (GWP) of CO_2, CH_4 and N_2O, which are 1, 23 and 296, respectively, and is based on net eCO_2 emissions from open dumping, land filling, aerobic composting and anaerobic digestion of OFMSW as well as ultimate use of compost as organic fertilizer and its impact on emissions and savings of eCO_2 achieved from avoidance of chemical fertilizer. The methodology stimulates recycling of organic waste. By aerobic composting and anaerobic digestion, organic matter originating from multiple waste streams is going through processes, which kills pathogens and produces compost which contains stable organic material which is useful for agricultural fields as a soil conditioner. Soil fertility, structure, water holding capacity and buffering capacity are all improved by this means.

Composting/digestion is acknowledged as an emission reduction project if the "baseline scenario" in a specific country causes significant greenhouse gas emissions (UNFCCC/CCNUCC 2008). This is for instance the case in Bangladesh, where most of the waste is land filled or illegally open dumped. When organic waste is land filled, a fermentation or rotting process will start due to a lack of oxygen. During fermentation, microbes will emit methane, a greenhouse gas which is 23 times stronger than carbon dioxide in term of GWP. The composting/digestion scenario "avoids" the emission of methane for a substantial part but on the other hand causes more emissions due to the transport of biomass and fuel use on the these facilities. The emissions of N2O due to microbial activities may also be higher during composting than during fermentation. However, the emission of N2O under anaerobic digestion is less as it is a closed process and no gas is released to atmosphere. There are many studies to quantify the GHG avoidance by aerobic composting and anaerobic digestion (Luske 2010; Brown et al. 2008; Shams 2017).

Morris et al. (2013) reported that anaerobic digestion and composting had similar negative GWP values by performing meta–analysis of 82 studies done on comparing aerobic composting and anaerobic digestion with different waste management systems. For composting and anaerobic digestion, the GWP was reported in the range -740 to -60 kg CO_2-eq. ton^{-1} initial wet weight (WW) for anaerobic digestion, while for aerobic composting, it was in the range -760 to 220 CO_2-eq. ton^{-1} initial WW. Menikpura and Sang-Arun (2013) concluded in the Simulation User Guide, version II (edited), that net GHG emissions from aerobic composting and anaerobic digestion are -1748.13 and -1048.94 kg CO_2-eq/ton of organic fraction of solid waste, respectively.

Based on the above reported negative GWP of aerobic composting and anaerobic digestion, it can be considered that 1 ton CO_2 eq/ton of organic fraction of solid waste could be reduced by applying any of these two processes. Based on the

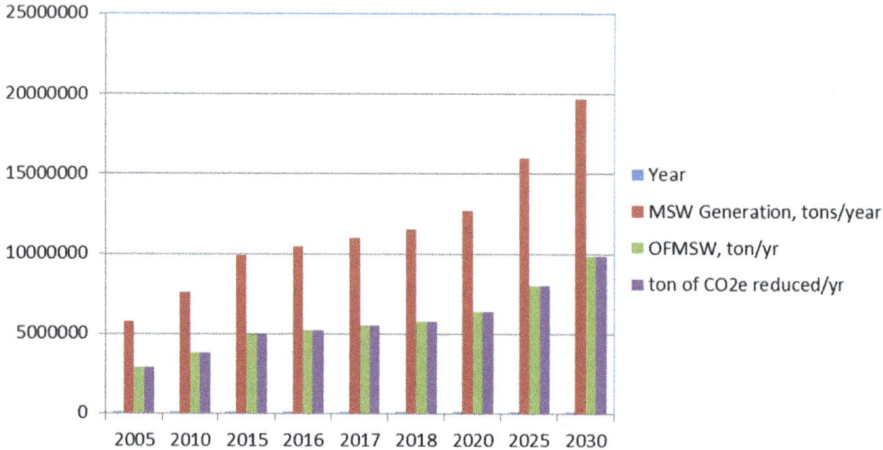

Fig. 8.5 Predicted reduction of GWP via aerobic composting or anaerobic digestion

quantity of waste generation in Bangladesh given in Table 8.4 and shown in Fig. 8.2 as well as life cycle analysis shown in Fig. 8.4, considering organic fraction to be 50% of the waste, reduction of GWP as shown in Fig. 8.5 would be substantial, which will play a vital role in climate change.

Issues to Be Considered for Improvement of MSWM in Bangladesh

To improve solid waste management in the urban areas of Bangladesh and to fulfill the desired life cycle shown in Fig. 8.4, the following issues should be considered:

- All the municipal authorities should promote aerobic composting and anaerobic digestion and discourage land filling of waste through paying gate fee to the composting or anaerobic digestion plants to treat MSW as are practiced globally as an avoidance cost of land filling.
- Promotion and grantee of source separation of waste through strong government regulations.
- Tax incentive for use and production of recycled product.
- Promotion of public–private–community partnerships.
- Ministry of Agriculture should bring change 15% moisture contain in the standard of organic fertilizer to minimum 45% to reduce the unnecessary drying cost of organic fertilizer produce from anaerobic digestion process as well as aerobic composting.
- Promotion of more waste-related projects using clean development mechanism (CDM) opportunity.

References

Ahsan M, Alamgir M, El-Sergany M, Shams S, Rowshon MK, Daud NNN (2014) Assessment of municipal solid waste management system in a developing country. Chin J Eng 11. Article ID 561935. https://doi.org/10.1155/2014/561935

Alamgir M, Ahsan A (2007a) Municipal solid waste and recovery potential: Bangladesh perspective. Iran J Environ Health Sci Eng 4(2):67–76

Alamgir M, Ahsan A (2007b) Characterization of MSW and nutrient contents of organic component in Bangladesh. Electron J Environ Agric Food Chem 6(4):1945–1956

Alamgir M, McDonald C, Roehi KE, Ahsan A (2005) Integrated management and safe disposal of MSW in least developed Asian countries—a feasibility study, WasteSafe. Khulna University of Engineering and Technology, AsiaPro Eco Programme of the European Commission

Andersen JK, Boldrin A, Samuelsson J, Christensen TH, Scheutz C (2010) Quantification of greenhouse gas emissions from windrow composting of gardenwaste. J Environ Qual 39:1–12

Baldé CP, Wang F, Kuehr R, Huisman J (2014) The global E-waste monitor, quantities, flows and resources, United Nations University, IAS—SCYCLE. Bonn, Germany

Bangladesh Health System Review (2015) Health systems in transition, vol 5(3)

Bari Q, Mahbub Hassan HK, Ehsanul Haque M (2012a) Solid waste recycling in Rajshahi city of Bangladesh. Waste Manag 32:2029–2036

Bari QH, Mahbub Hassan K, Ehsanul Haque M (2012b) Scenario of solid waste reuse in Khulna city of Bangladesh. Waste Manag 32:2526–2534

BCAS (1998) Refuse quantity assessment of Dhaka citycorporation for waste to electrical energy project, finalreport. Bangladesh Centre for Advanced Studies, World Bank, GOB, Dhaka, Bangladesh

BCSIR (1998) Refuse quality assessment of Dhaka city corporation for waste to energy project, institute of fuelresearch and development. Bangladesh Council of Scientificand Industrial Research, World Bank, GOB, July, Dhaka, Bangladesh

Bhide AH (1990) Solid waste management at Dhaka, Khulna and Natore. WHO Project BAN CWS 001: Detailed Technical Report

Bhuiya GMJA (2007) Bangladesh. Solid waste management: issues and challenges in Asia. The Asian Productivity Organization (APO), Tokyo, Japan, p 29

Bhuiyan MSH (1999) Solid waste management of Dhakacity. Paper presented at the seminar on solid waste management, DCC, Dhaka, Bangladesh

BKH (1985–86) Housing development project-subcontract "A", Annexure-V- solid waste management. The BKH Consultant, Delft, The Netherlands., UNDP, UNCHS (Habitat)

BMDF (2012) Study on municipal solid waste management, Final Report, Bangladesh Municipal Development Fund (BMDF), 21 June 2012

Brown S, Kruger C, Subler S (2008) Greenhouse gas balance for composting operations. J Environ Qual 37:1396–1410

BUET (2000) Characterization of municipal solid waste and preliminary environmental impact assessment of collection and disposal works in Dhaka city'. Department of Civil Engineering, Bangladesh University of Engineering andTechnology, BRTC, DCC

CCAP (2013) Dhaka's integrated municipal solid waste program, CCAP-Booklet Bangladesh, Center for Clean Air Policy, Washington, DC, http://ccap.org/assets/CCAPBooklet_Bangladesh.pdf. Retrieved on 15 Jan 2015

Chowdhury DAH, Mohammad N, Haque MRU, Hossain DT (2014) Developing 3Rs (reduce, reuse and recycle) strategy for waste management in the urban areas of Bangladesh: socioeconomic and climate adoption mitigation option. IOSR J Environ Sci Toxicol Food Technol (IOSR-JESTFT) 8(5):09–18

DCC (1985) Dhaka city corporation, Unpublished Report

Diaz LF, Savage GM, Eggerth LL (1993) Composting and recycling municipal solid waste. CalRecovery Inc, USA

DNCC (2014a) Dhaka North City Corporation, www.dncc.gov.bd/dncc-setup/ chronologicaldevelopment-of-dncc.html. Retrieved on 25 July 2014

DNCC (2014b) Dhaka North City Corporation. http://www.dncc.gov.bd/departments-withfunction/department-s/solid-waste.html. Retrieved on 25 July 2014

DNCC (2015) Dhaka North City Corporation, http://www.dncc.gov.bd/departments-withfunction/ department-s/solid-waste/solid-waste-2.html. Retrieved on 25 Mar 2015

DNCC Waste Report 2016–2017. http://dncc.portal.gov.bd/sites/default/files/files/dncc.portal.gov. bd/notices/6fbdcf34_b55a_4fd2_887f_6e78d5ada3c5/Waste%20Report%202016-2017.pdf

DoE (2013) Bangladesh Environment and Climate Change Outlook (ECCO) 2012, Department of Environment, Ministry of Environment and Forests, Government of the People's Republic of Bangladesh, Dhaka, published in June 2013

Enayetullah I, Khan SSA, Sinha AHMM (2005) Urban solid waste management. Scenario of Bangladesh: problems and prospects. Waste concern technical documentation. Waste Concern, Dhaka, Bangladesh

Forbes (2013) http://www.forbes.com/fdc/welcome_mjx.shtml. Retrieved on 22 June 2013

Hai FI, Ali MA (2005) A study on solid waste management system of Dhaka City corporation: effect of composting and landfill location. UAP J Civil Environ Eng 1(1):2005

Harun AHMA (2017) Engineer, Waste Management Department, DSCC, personal communication, Data sent via e-mail on 13 Nov 2017

Hossen MM, Rahman AHMS, Kabir AS, Faruque Hasan MM, Ahmed S (2017) Systematic assessment of the availability and utilization potential of biomass in Bangladesh. Renew Sustain Energy Rev 67:94–105

Islam AZMS (2003) Solid waste management in Dhaka City, Institute for Global Environmental Strategies, Kitakyushu Initiative for a Clean Environment. http://kitakyushu.iges.or.jp/docs/ mtgs/seminars/theme/swm/presentation/3%20Dhaka%20%28Paper.pdf. Retrieved on 12 Aug 2014

Islam KM (2017) Addl. Chief Waste Management Officer, DSCC, personal communication on 14 Nov 2017 at 9.30 hrs

Islam MS, Pervin L, Muyeed AA (2012) Feasibility study for using the water ways to transport of solid waste of Dhaka City and waste transport route preparation using GIS for DCC solid waste management. Can J Environ Constr Civil Eng 3(1)

JICA (2005) The study on the solid waste management in Dhaka City. Clean Dhaka Master Plan Project, Japan International Cooperation Agency, Pacific Consultants International and Yachiyo Engineering Co. Ltd. DCC, Dhaka, Final Report, Volume 2, March 2005

JICA (2012) Summary of terminal evaluation. Clean Dhaka Master Plan Project, Japan International Cooperation Agency. http://www.jica.go.jp/english/our_work/evaluation/tech_ and_grant/project/term/asia/c8h0vm000001rr8t-att/bangladesh_2012_01.pdf. Retrieved on 30 Mar 2015

LBI (1990) Dhaka integrated flood protection project (FAPbB). Louis Berger International, ADB

Lukse B (2010) Reduced GHG emissions due to compost production and compost use in Egypt, Comparing two scenarios. Soil and More International BV, Louis Bolk Institute

Menikpura N, Sang-Arun J (2013) "User manual: estimation tool for greenhouse gas (GHG) emissions from municipal solid waste (MSW) management in a life cycle perspective", simulation user guide, version II (edited). Institute for Global Environmental Strategies, Japan

MMI (1991) Dhaka metropolitan development planning (DMDP) waste management report', Mott Macdonald Int. Ltd. and Culpin Planning Ltd., UNDP, UNCHS, GOB, RAJUK

Morris J, Matthews HS, Morawski C (2013) Review and meta-analysis of 82 studies on end-of-life management methods for source separatedorganics. Waste Manag 33(3):545–551

PAS (1997) Integrated waste management programme for Dhaka City corporation', waste to energy plant, composting and multiple thermal by-products. Pan AsiaServices Ltd., Dhaka and CGCA-ONYX, France

PCI (1991) Greater Dhaka protection project. Pacific Consultants Internationals International, Japan

Rahman MA (2017) E-waste management: a study on legal framework and institutional preparedness in Bangladesh. Cost Manag 45(1):28–35. ISSN 1817-5090

Rahman SMS, Shams S, Mahmud K (2013) Study of solid waste management and its impact on climate change: a case study of Dhaka City in Bangladesh. Retrieve from http://benjapan.org/iceab10/62.pdf on 15 April 2014

RSWC (1998) Waste landfilling and hospital waste incineration in Dhaka ROTEB-Solid Waste Consultancy. B.V. Netherlands

Shams S, Sahu JN, Rahman SMS, Ahsan A (2017) Sustainable waste management policy in Bangladesh for reduction of greenhouse gases. Sustain Cities Soc 33:18–26

Tchobanoglous G, Theiswn H, Vigil S (1993) Integrated solid waste management. McGraw-Hill International Editions, Civil Engineering Series, Network, U.S.A

UNCRD (1998) Compendium of facts and figures on solid waste management in Asian metropolises, prepared for UNCRD Research Project RES/642/87 on improving solid waste management in the context of Metropolitan Development and Management in Asian Cities

UNESCAP (2009) Statistical yearbook for Asia and the pacific 2009. http://www.unescap.org/stat/data/syb2009/2-Urbanization.asp,retrieved on 5 July 2013

UN-HABITAT (2013) Urban planning for city leaders. United Nations Human Settlement Programme (UN-Habitat), Nairobi, Kenya, ISBN: 978-92-1-132636-9, December 2013

Wang H, Nie Y (2001) Municipal solid waste characteristics and management in China, technical paper. Air Waste Manag Assoc 51:250–263

Waste Atlas. Country Data: Bangladesh. [cited 2014; Available from: www.atlas.d-waste.com]

Waste Concern (2005) Urban solid waste management scenario of Bangladesh: problems and prospects, technical documentation, waste concern, Bangladesh

Worldometers (2017) Retrieved on November 8, 2017 from www.Worldometers.info, which uses population data from UN

Yousuf TB (2005) Sustainability and replication of community-based composting-a case study of Bangladesh. PhD thesis, Loughborough University, UK

Yousuf TB, Rahman M (2007) Monitoring quantity and characteristics of municipal solid waste in Dhaka City. Environ Monit Assess 135:3–11

Chapter 9
Impacts and Responses of Bangladesh Coastal Fishing Communities to Climate Change: Implications for Policy and Scaling-up

Md. Monirul Islam

Abstract Climate change is predicted to impact on fisheries and dependent communities. This study assesses the vulnerability and adaptation to the impacts of climate variability and change, in three small-scale coastal fishing communities in Bangladesh with a view to suggest policy and scaling up the findings. Overall, using a mixed method approach this study contributes empirical evidence to current debates in the literature on climate change by enhancing an understanding of the characteristics and determinants of livelihood vulnerability, migration as an adaptation strategy and limits and barriers to the adaptation of fishing communities to climate variability and change. This study finds that the coastal fishing communities have been impacted by several climatic shocks and stresses and they have traditionally coped with or adapted to the normal range of climate impacts but not always sufficiently well. This study suggests that reduction of impacts, vulnerability or risks, increase in adaptive capacity or resilience, and facilitating adaptation actions and processes to climate variability and change for the fishing communities would require multifaceted measures. However, caution should be maintained as some adaptation strategies may exacerbate existing problems or may be maladaptive to other systems. The findings of this study would particularly contribute to the Government of Bangladesh's policy goal of "assess[ing] potential threats [of climate change] to the marine fish[eries] sector and develop[ing] adaptive measures". The findings of this study may also *partly* be transferred and scaled up to other coastal fishing communities in the Bay of Bengal region with similar socio-economic and environmental characteristics.

Keywords Vulnerability · Adaptation · Fishing communities · Climate policy · Bangladesh

Md. M. Islam (✉)
Department of Fisheries, University of Dhaka, Dhaka 1000, Bangladesh

© Springer Nature Switzerland AG 2021
Md. Jakariya and Md. N. Islam (eds.), *Climate Change in Bangladesh*,
Springer Climate, https://doi.org/10.1007/978-3-030-75825-7_9

157

Introduction

Climate variability used to be a normal phenomenon in the Earth's history, but over the last few decades climate has been changing faster and is predicted to do so even more in coming decades due to global warming (Stocker 2014). This faster climate change has been predicted to impact on both natural and human systems in a complex and unprecedented way (Field et al. 2014). The fishery sector, which supports livelihoods of 660–820 million people (FAO 2012), is considered among the worst affected by climate change (Field et al. 2014). Climate change is an additional pressure on fishery systems which already experience other stresses such as overfishing, loss of habitat, pollution and disturbance (Brander 2006; Coulthard 2009). In particular, small-scale fishing communities in developing countries, which constitute 90% fishery-dependent people (FAO 2012), will face complex and localised impacts, as predicted by the Intergovernmental Panel on Climate Change (IPCC) with high confidence (Climate Change 2007a). The IPCC predicts with high confidence that climate change will cause marine species redistribution and marine biodiversity reduction which will challenge the sustained provision of fishery productivity (Field et al. 2014). Other impacts include damage in fishery methods, and land-based property and infrastructure (Westlund et al. 2007; FAO 2008). These have the potential to make fishing communities and their livelihoods more vulnerable, but they are only occasionally investigated in the context of developing countries or investigated in other sectors such as agriculture. A detailed study on how fishing communities are vulnerable to past and current climate impacts can provide important insights to address the enhanced level of future impacts or reduce vulnerability for them.

To address the impacts of climate change, adaptation is widely recognised as an important response strategy along with mitigation (Fankhauser 1996; Smith 1996; Adger et al. 2007). However, due to lag times in the climate and biophysical systems, the positive impacts of current mitigation efforts will not necessarily be noticeable until around 2050 (Climate Change 2007b). The current level of greenhouse gases will continue to change the climate in the next few decades (Climate Change 2007b). Therefore, adaptation is regarded as inevitable and necessary to tackle the additional shocks and stresses due to climate change (Pielke et al. 2007; Stern 2007). While societies, including fishing communities, have traditionally adapted to the normal range of climatic variation using different strategies, this level of adaptation is not distributed homogeneously around the world (Perry et al. 2010). Climate change is predicted to pose impacts and vulnerability often outside the range of experience, for which additional adaptation will be needed (Adger et al. 2003), especially for fishing communities (FAO 2008; Allison et al. 2005). As such, fishing communities deserve greater attention within climate change adaptation debates because they face compounding climate change impacts and non-climatic pressures (Coulthard 2009).

Human migration is regarded as one of the strategies to cope with or adapt to the impacts of climate change. This strategy has brought much attention in recent years

as it is predicted that millions of people, many of whom are from fishing communities, are likely to be displaced due to the impacts of climate change (Myers
2002; Nicholls et al. 2011). However, there is an ongoing debate in academic and
policy arenas about the successfulness of climate-induced migration. Migration
may be short term (temporary/seasonal) or long term (permanent), short distance
(internal) or long distance (international), and forced (reactive) or voluntary
(adaptive). While a growing body of literature considers different drivers
(McLeman and Smit 2006; Black et al. 2011a, b; McLeman 2011; Piguet et al.
2011; GOS 2011) and types of migration (Piguet et al. 2011; Paavola 2008), only a
few of them examine the likely consequences of migration (Black et al. 2011b;
GOS 2011; Paavola 2008; Mortreux and Barnett 2009; Barnett and O'Neill 2012).
None of them have used evidence-based data to conclude the outcomes of migration, and many studies have asked for more empirical studies on this issue to
support public policy (e.g. IPCC 2007a; Stern 2007; GOS 2011). Studies on the
outcomes of past and present climate-induced migration can therefore provide
important insights for developing strategies to cope with and adapt to climate
change. Especially, comparing a fishing community that migrates permanently due
to climatic reasons, leaving a portion behind, provides an opportunity to compare
the two communities and assess the successfulness of migration.

Fishing communities can adapt in many ways, and migration is just one
example. Adaptation efforts are impeded in many ways, however. Limits (largely
insurmountable constraints) and barriers (often malleable constraints) can constrain
people's ability to identify, assess and manage risks in a way that maximises their
well-being and facilitates adaptation to climate variability and change (Climate
Change 2007a; IPCC 2012; Adger et al. 2009; Moser and Ekstrom 2010). Fishing
communities may not be an exception in this respect (Morgan 2011). Many of these
limits and barriers are interrelated and combine to constrain adaptation (Adger et al.
2007; Jones and Boyd 2011). But there is a lack of evidence on limits and barriers
to adaptation and interactions between them, especially from a developing country
perspective. Assessing these limits and barriers would help find suitable means of
overcoming them to enable the adaptation of fishing communities to present-day
climate variability and future climate change.

Bangladesh is regarded as one of the most vulnerable countries to the impacts of
climate change (Climate Change 2007a; Met Office 2011; World Bank 2013a),
despite its significant economic strides over the past four decades (World Bank
2013b). Its fishery sector, regarded among the most vulnerable to climate change in
the world (Allison et al. 2009), supports the livelihoods of about 7 million fishers
directly and contributes 4.43% to GDP and 2.73% to export earnings (DoF 2012).
Most (93%) of the marine fishing is small scale in nature, supporting livelihoods of
over half a million fishers and their household members (ibid). The climate of
Bangladesh has changed over the past decades, and predictions are that it will
continue to change even more in future, resulting in considerable negative impacts
especially in the coastal areas (Met Office 2011). From 1980 to 2000, a total of
250,000 deaths were associated with tropical cyclones around the world, of which
60% occurred in coastal Bangladesh (Climate Change 2007a). One of the most

devastating cyclones and associated storm-surge-induced floods killed 300,000 people in coastal Bangladesh in 1970 (Climate Change 2007a), many of whom were from fishing communities. Thus, the coastal fishing communities in Bangladesh are particularly interesting cases for the study of vulnerability and adaptation to climate variability and change. The findings of such a study could also contribute to an understanding of these issues in other parts of the world with similar environmental, socio-economic and livelihood conditions.

In summary, despite considerable studies on the impact of climate change on aquatic ecosystems and fish populations, on macro-scale fishery-dependent economies and their people, and on vulnerability and adaptation in agricultural communities, there has not yet been sufficient examination of the vulnerability and adaptation of small-scale fishing communities to climate variability and change. This study aims to contribute to an increased understanding of these issues and, in particular, of the situation facing Bangladesh. Overall, based on empirical evidence this study contributes to current debates on climate change by enhancing an understanding of the characteristics and determinants of livelihood vulnerability, migration as an adaptation strategy and the limits and barriers to adaptation of fishing communities to climate variability and change. These have important implications for policy.

After this introduction, Sections "Climate Variability and Change in Bangladesh"–"Coastal Fishing Communities and Climate Change in Bangladesh" review the climate change, fisheries and policy in Bangladesh (Sections "Climate Variability and Change in Bangladesh"–"Coastal Fishing Communities and Climate Change in Bangladesh"). The methodology is described in Section "Case Study, Materials and Methods". Based on the empirical findings of three case studies (Section "Policy Implications"), this study suggests policy implications (Section "Policy Implications") and scaling up the findings (Section "Scaling-up and Transferability of Findings"). Section "Results and Discussion" is based on the three studies done by Islam et al. (2014a, b, c)

Climate Variability and Change in Bangladesh

According to the Met Office (2011), Bangladesh has experienced widespread warming (0.24 °C per decade during the hot season of March to May and 0.19 °C per decade during the cool season of December to February) and a small increase in total precipitation since 1960. This study also observed several temperature and precipitation extreme events in the last 5 decades. Although this study observed a long-term trend of temperature extremes, no evidence of a long-term trend of precipitation extremes (i.e. continuous wet or dry days) was observed. Between the years 1985 and 2009, an increased rate of sea surface temperature (0.0086–0.0191 °C annually) was found in the Bay of Bengal (Chowdhury et al. 2012). Other studies (Mirza and Dixit 1997; Khan et al. 2000; Mirza 2002) found an increase in temperature of about 1 °C in May and 0.5 °C in November (from 1985 to 1998), and decadal rain anomalies above long-term averages since the 1960s in Bangladesh. Shahid (2010) observed an increase in annual and pre-monsoon rainfall as 5.53 and 2.47 mm/year, respectively, over the period 1958–2007.

In coming decades, Bangladesh will experience a mixture of climate variability and change (some data are shown in Table 9.1). Greater variation in temperature and precipitation has been predicted compared to the past. General Circulation Models for Bangladesh predict a steady increase in temperature and precipitation (Agrawala et al. Agrawala et al. 2003a), with temperatures increasing in both winter and summer, but more so in winter. In contrast, precipitation is predicted to decrease in winter and much increase in summer (Agrawala et al. 2003a). The Met Office (2011) projects 3–3.5 °C increase in temperature in Bangladesh, 20% increase in precipitation in the north of the country and 5–10% increase in precipitation for the rest of the country by 2100 under A1B (higher) emission scenario of IPCC.

About 7% of the global cyclonic storms are formed in the Bay of Bengal region which is considered a potentially energetic region for the development of cyclonic storms (Gray 1968). Between the years 1985 and 2009, an increased frequency of cyclonic storms (5.48 annually) was experienced in the Bay of Bengal (Chowdhury et al. 2012). There is still considerable uncertainty regarding how cyclone frequency and intensity will be affected in future by climate change in Bangladesh (Met Office 2011). Cyclone frequency may increase (McDonald et al. 2005; Sugi et al. 2009) or decrease (Bengtsson et al. 2007; Emanuel et al. 2008; Zhao et al. 2009) in the North Indian Ocean (home to the Bay of Bengal). Likewise, cyclone intensity may increase (Vecchi and Soden 2007; Yu et al. 2010a) or decrease (Oouchi et al. 2006) or even remain unchanged (Emanuel et al. 2008) in this basin. A more recent study, specifically focussing on the Bay of Bengal, predicted an increasing frequency of 7.94 cyclonic storms per year by 2050 due to climate change (Chowdhury et al. 2012).

Along the Bangladeshi coast, the sea level has risen in the past and is predicted to rise further in future. Sea level rise is influenced by astronomical, geological and climate change factors. Expansion of the ocean, due to the increased temperature and the melting of glaciers, small icecaps, the Greenland ice sheet and the Antarctic ice sheet, is the major climate-related factor that could explain a rise in global mean sea level on a 100-year timescale (Warrick and Oerlemans 1990). Yu et al. (2010b) observed that the sea level has risen along the Bangladeshi coast over the past 60 years. Depending on the coastal geomorphology, it varied along the coastline. At Hiron point (western region), the rate of sea level rise was 5.6 mm/year, whereas at the Chandpur station (central region) on the Meghna River no change was observed. Although the sea level rose at most of the stations, only that at Hiron point was statistically significant. By 2030, 2050 and 2100, the sea levels along the Bangladeshi coasts may raise up to 14, 32, and 88 cm, respectively (MoEF 2005). In a review, Agrawala et al. (2003a) predicted the sea level rise in Bangladesh as 30 cm–1 m by 2100 under different IPCC scenarios.

An increase in temperatures, sea level and the number of summer precipitation events and cyclones will result in higher intensity and frequency of

Table 9.1 Future climate change scenarios for Bangladesh ([1]Agrawala et al. 2003a; [2]Emanuel 1987; [3]MoEF 2005; [4]Mirza 2003)

Year	Mean temperature change (°C)[1]			Mean precipitation change (%)[1]			Cyclone (% increase of wind speed)[2]	Sea level rise (cm)[3]	Flood (% increase of flooded area)[4]
	Annual	December, January, February	June, July, August	Annual	December, January, February	June, July, August			
Baseline average (mm)				2278	33.7	1343.7			
2030	+1.0	+1.1	+0.8	+3.8	−1.2	+4.7		14	
2050	+1.4	+1.6	+1.1	+5.6	−1.7	+6.8		32	
2100	+2.4	+2.7	+1.9	+9.7	−3.0	+11.8	10–25	88	23–29

storm-surge-induced floods in coastal Bangladesh. Of the total land area, 79% has less than 1 m elevation that includes all the coastal areas (Rashid 1991). With a global temperature rise of 2 °C, the flooded area in Bangladesh will rise by at least 23–29% more than today (Mirza 2003). The flooded area, flooding depth and surge intrusion length may be substantially larger under intensified surge conditions (Karim and Mimura 2008). As Dasgupta et al. (2011) suggested, a 10% intensification of the storm surge combined with a 1 m sea level rise could affect 23% of Bangladesh's total coastal land area.

Land erosion is a regular and recurring phenomenon in Bangladesh. However, in coastal areas, especially on small islands, the rate of erosion is higher. A comparison of Landsat imagery of 1972 and 1987 showed a total of 11 small Bangladeshi coastal islands (and/or chars) disappeared totally (Pramanik 1988). Although this study does not specifically mention the number of islands that might have been accreted, it has found that during that period erosion was greater than accretion, resulting in a net loss of the number of islands. Of the total 50 islands in 1973, only 39 existed in 1987 (Pramanik 1988). Under sea level rise and increased flooding, land erosion is predicted to intensify along Bangladeshi coasts if protection is not given (Ahmed et al. 2002). The climate change scenario is predicted to increase the volume of water in the Ganges–Brahmaputra–Meghna river system during the monsoon. This may also increase coastal land erosion (Agrawala et al. 2003b).

The coastal region of Bangladesh is susceptible to increasing salinity in groundwater as well as surface water resources. Referring to the available data of 2005, Yu et al. (2010b) found that about 12% land area of Bangladesh contains a salinity of more than 5 ppt during the monsoon which goes up to 29% during the dry season. The sea level rise (Han et al. 1999) and increased storm surge (Ahmed et al. 2002) are the two reasons for this increased level of salinity. Thus, increased levels of sea and storm surge will result in more intrusion of salinity in the coastal areas.

In addition to model-based observations/predictions mentioned above, local elderly people in coastal Bangladesh observed a continuous shift in climatic patterns, timing of the onset of monsoon and the highest level of tidal levels (Rahman et al. 2007).

Coastal Fishing Communities in Bangladesh

Bangladeshis have a long tradition of fishing and fish culture which contribute significantly to employment, income generation, export earnings and human nutrition. This sector supports livelihoods of about 7 million fishers directly and 12 million people indirectly and contributes 4.43% to GDP and 2.73% to export earnings (DoF 2012). Most (93%) of the marine fishing is small scale in nature and supports the livelihoods of over half a million fishers and their household members (DoF 2012) living in 870 fishing communities (Aghazadeh 1994). Although no recent data are available, the number of coastal fishing communities is frequently claimed as more than 2000 in the media. In addition, marine fisheries support the livelihoods of other households involved in ancillary activities such as fish processing, gear making and so on.

Several studies have found poor physical infrastructure in the coastal fishing villages of Bangladesh, and most people live in poor socio-economic conditions (Ahmed et al. 2009; BOBP 1985; Ahmed 2002; Chowdhury 2009; Hasan et al. 2004; Akter et al. 2009). They have also found that most of the households cannot eat regularly, have little education, and have only moderate public health provision. Some get financial assistance from the government and international donors (Hasan et al. 2004). Local village leaders tend to make community decisions and resolve most family conflicts, although sometimes elected local government representatives such as the chairmen and members of *"Union Parishad"* (a local government unit) resolve conflicts (Ahmed et al. 2009). Women have less freedom both socially and economically than men (Ahmed et al. 2009). However, most of them can cast their votes during national and local government elections (Ahmed et al. 2009).

In the fishery sector, women are mainly involved in post-harvest, processing and marketing. It is estimated that about 30% of Bangladeshi women in rural coastal areas are directly or indirectly engaged in small-scale fishing activities (Libreo 1987). Their specific activities include making and repairing fishing gear, sorting fingerlings (especially in coastal areas), catching shrimp/prawn post-larvae, fish processing, transportation, small-scale marketing (Ahmed et al. 2012). However, they are often excluded from fishing and from the institutions that manage fisheries (Sultana and Thompson 2006).

Marine small-scale fishery-dependent people can be categorised into different groups (Table 9.2). Most of the fishers catch fish with boats and gear, although a small number of them do not have a boat and operate only with small (push/pull) nets near the shore. Four types of boats (small manual, small mechanised, medium mechanised and large mechanised) are normally used with different types of nets depending on the target species and fishing season. In Bangladesh, there are 21,097 total motorised fishing vessels of which 99.20% are less than 12 m in length (FAO 2012). Normally rich people, who can afford at least 400,000T K, own these boats (BOBP 1985). In a boat, a group of 5–25 people work during a fishing operation that lasts between 12 h and 20 days (Hasan et al. 2004). In a typical large boat, there is a captain (also called crew leader) and normal crew members (BOBP 1985). The boat captain's income is two to three times higher than a normal crew member (Hasan et al. 2004).

There are two main fishing seasons: rainy and winter. In the rainy season (May to September) with rough seas, mainly Hilsa shad fish (*Tenualosa ilisha*) (the most popular and the largest single species fishery of Bangladesh) are caught by gill nets on the far shore. Some other species such as Bombay duck (*Harpadon nehereus*), pomfret (*Pampus* spp), ribbon fish (*Trichiurus savala*), flat fish and shrimps (*Penaeus* spp) are also caught during this season by using bag nets, longlines and trammel nets. In the winter season (November to March), the sea generally stays calm. During this season, some of the above species of fishes are caught but shrimp

Table 9.2 Classification of coastal and marine small-scale fisheries in Bangladesh

Characteristics	Classifications
Activity	Boat owning and renting (usually males)
	Fishing (as crew leader or captain) (only males)
	Fishing (as crew member) (only males)
	Shrimp post-larvae collecting (mainly females, elderly and children)
	Fish processing (both males and females)
	Fish trading (mainly males; some females)
	Boats and gear making and mending (both males and females)
Type of boats	Small manual boat
	Small mechanised boat
	Medium mechanised boat
	Large mechanised boat
Type of gear	Gill net
	Bag net
	Longline
	Trammel net
	Others
Scale of fishing	Full-time
	Part-time
	Occasional

Source BOBP (1985), Hasan et al. (2004), Akter et al. (2009), DoF (2009)

post-larvae are also collected during this season by small manual boats with bag nets and by push/pull nets without a boat. During the other two months of the year (April and October), the boats and nets are repaired and the fishers prepare for fishing (Hasan et al. 2004). There are three types of fishers depending on time involvement in fishing—full-time (9–12 months per annum), part-time (3–9 months per annum) and occasional (less than 3 months per annum). After landing, the fish are normally sold by auction in local fish landing centres or on the local market or directly to local fish processors. Fish marketing is controlled by a group of intermediaries known as "*Aratdars*" (commissioning agents) and "*Mohajans*" (money lenders). The commissioning agents dominate the wholesale markets and have a chain of suppliers who regularly bring catches. These agents charge 3–6% commission and take 2–4 fish for every 80 fish sold (Rahman 1994). The agents in turn provide advance money *(dadon)* to boat owners to make boats and nets. The boat owners are required to sell fish to the agents. After landing, fishermen tend to sell their fish as early as possible to these agents to avoid spoilage because of the inadequate cold storage facilities and unavailability of good quality ice (Ahmmed 2007). There are around 6,500 fish markets scattered across the country of which 4,500 are small primary village markets (Rahman 1994).

Seafood is processed and preserved in mainly two ways: freezing (includes chilling) and drying (includes salting followed by drying). There are 162 shrimp and fish processing (freezing) industries (DoF 2012). Fish are frozen as whole (often beheaded and gutted) or as fillets or steaks, by either individual quick frozen or block frozen technique (Ahmmed 2007). Fish that are exported overseas need to achieve international standards following international regulations such as the Hazard Analysis Critical Control Point (HACCP) procedure.

In Bangladesh, traditional sun drying is one of the most popular low-cost methods of fish processing and preservation for both domestic consumption and export. Traditional sun drying is carried out in the open air using solar energy to evaporate the water from fish. This technique involves a longer drying period, no control over the operating variables and a risk of infestation with insects, their eggs and larvae (Islam et al. 2007, 2008). To overcome these disadvantages, a small quantity of fish is also dried by mechanical drier such as a solar tunnel drier where different variables such as temperature, relative humidity, air velocity and solar radiation are controlled (Islam et al. 2007, 2008; Bala and Mondol 2001).

Coastal Fishing Communities and Climate Change in Bangladesh

There are few studies available on climate change and the fisheries of Bangladesh. A national scale study found that the economy of Bangladesh will be among the most vulnerable to climate change impacts on fisheries by the 2050s (Allison et al. 2009). The results from this type of study should however be used cautiously as it has used only a single climate change parameter (Met Office 2011). The projected climate change may directly impact on the fish stocks and the Bay of Bengal ecosystems, and on the livelihoods of the fishery-dependent people in Bangladesh. In general, the impacts of climate change between the Bangladeshi coastal small-scale fishing communities and those of other parts of the world may have some level of similarity as the nature of this fishery system varies little across the world. Climate change may result in an increased level of fluctuation in fish production in Bangladesh (Ahmed et al. 2002; Ali 1999). Cyclones and associated floods may exert tremendous impacts on fishing assets and infrastructure and ultimately on the livelihoods of fishing communities. Ahmed and Neelormi (2008) observed a reduction of fishing days during the monsoon of 2007 due to cyclonic sea condition. More frequent and intensified cyclones can further reduce fishing days. In coastal Bangladesh, cyclones of very high intensity may occur in April and May, and between September and November (Met Office 2011). Most of these months fall within the fishing seasons, and consequently fishing activities may be impacted by intense cyclones. Traditional fish drying activity may also be impacted by increased temperature and variation in rainfall as well as by extreme climate and weather events. Sea level rise and land erosion may make the current living areas of fishing communities unsuitable and may result in their displacement or may leave them in a more vulnerable situation. As a whole, they are likely to be exposed more to climate change impacts (Agrawala et al. 2003a).

The Government of Bangladesh is claimed to be one of the pioneers and key proactive policy formulators in the world in addressing the negative impact of climate change in Bangladesh (MoEF 2012). Two main policy documents—the National Adaptation Programme of Action (NAPA) developed in 2005 (updated in 2009) and Bangladesh Climate Change Strategy and Action Plan (BCCSAP) developed in 2008—are frequently referred to in support of this claim (MoEF 2005, 2009; BCCSAP 2008). The BCCSAP builds on the NAPA. These documents provide overall policy and action guidance to address the impacts of climate change. The BCCSAP is the main basis of a ten-year plan to address climate change over the next 20–25 years. The objective of BCCSAP is to increase the country's resilience to climate change, reduce the climate change risks to national development and accelerate the development of the country following a low-carbon path. It also suggests an integrated approach to socio-economic development and the management of climate change issues. It has already received funding from international donors and has itself funded the implementation of some of the strategies.

However, how these strategies have helped or would help the coastal fishing communities to address climate impacts is vague. The lack of policy for coastal fishing communities is due to the serious lack of understanding of the issues and implications of climate change among scientists, activists and planners (Rahman et al. 2007). The first version of NAPA had two policies directly relevant to fisheries, consisting only of policies linked to culture fisheries but there was no policy for coastal and marine capture fisheries on which millions depend. The updated version of NAPA and especially the BCCSAP included two medium- and long-term strategies and action plans for adaptation in the coastal and marine fishery sector: (1) "Assess potential threats to fish spawning and growth of fish in the coastal zone and brackish water and develop appropriate adaptive measures and cultural practices", and (2) "Assess potential threats to the marine fish sector and develop adaptive measures". This research has the potential to contribute to the 2nd strategy.

Case Study, Materials and Methods

Selection of the Fishing Communities

In order to achieve the aim and objectives of this research, three coastal fishing communities (cases) were chosen in Bangladesh (Fig. 9.1). The communities were not selected randomly but rather based on a well-defined theoretical focus which replicates or extends theory by filling conceptual categories (Eisenhardt 1989; Yin 1984) and which has the potential for policy implications.

Due to the lack of accessible and published local-level information, it was challenging to select the fishing communities. The selection process comprised several steps including a literature review and a reconnaissance study. Through a literature review, two districts, Barguna and Cox's Bazar, were selected from the entire coastal area of Bangladesh. These districts are more affected by climatic

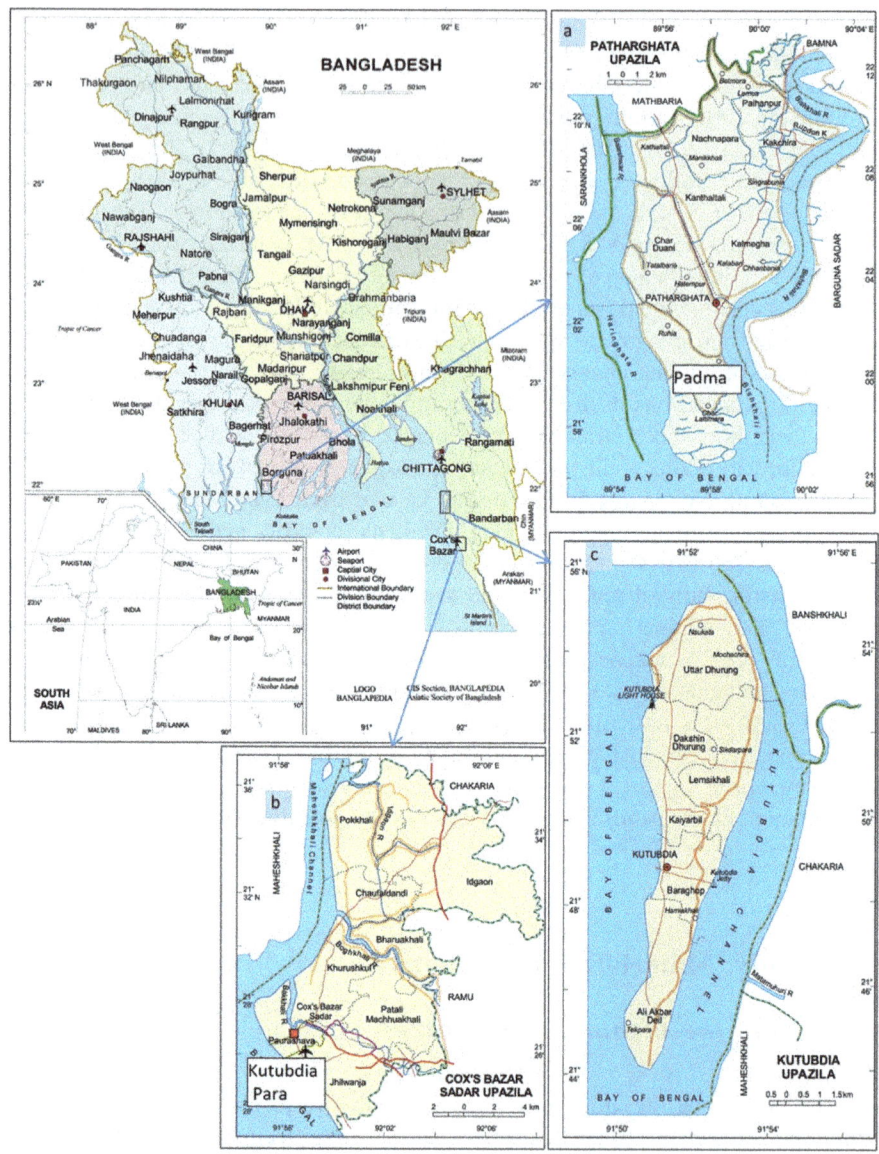

Fig. 9.1 Map of study sites in Bangladesh: **a** Padma within Patharghata sub-district, **b** Kutubdia Para within Cox's Bazar sub-district and **c** Kutubdia Island

phenomena such as cyclones and tidal fluctuation than other coastal areas of Bangladesh (Agrawala et al. 2003a). During the reconnaissance study, key information on climate and the livelihoods of the fishing communities were gathered using key informant interviews and focus group discussions (FGDs). Some of the local-level documents (such as demographics and past extreme events data), which are not available online, were also collected during that time. Three study sites were selected (Padma, Kutubdia Para and Kutubdia Island) (Fig. 9.1). The main selection criteria were characteristics of the settlements, communities' level of dependence on marine fisheries, their livelihood characteristics and their exposure to past climatic shocks and stresses (Table 9.3).

Data Collection and Analysis

This research has assessed the vulnerability and adaptation of three coastal fishing communities to the impact of climate variability and change in Bangladesh. The case study approach has combined climate change vulnerability and adaptation approaches, composite index approach and sustainable livelihood approach (SLA) to develop the research framework and guide the overall data collection and analysis processes.

This study follows a mixed method approach of data collection (Tables 9.4 and 9.5). Quantitative methods (e.g. structured household questionnaires) were used mainly for collecting data on context, whereas qualitative methods (e.g. oral history interviews, vulnerability matrices, key informants interviews and FGDs) were used to get rich, detailed and contextually grounded data (Nightingale 2003).

Data were collected only from fishery-dependent households across the communities, except from some key informants outside the communities (described later in Sect. 3.3.5) who are directly associated with the communities' interests. Fishery-dependent households include any level of, full-time, part-time or occasional, dependence on fisheries such as fishing, fish processing, boats and gear making and mending, and fish trading. A sampling frame for the fishery-dependent households was prepared in each community before data collection. Of the total 908 households in Padma, 89% (811 households) are fishery-dependent. Of the total 1193 households in Middle and North Kutubdia Para (this study was conducted in these two sections of Kutubdia Para), 83% (994 households) are fishery-dependent. From Kutubdia Island, 89 households were targeted—11 households that remain in Kuzier Tek and another 78 who are settled on other parts of the island. All of them are fishery-dependent at least occasionally.

The data were collected in October 2010 (reconnaissance study) and between February and July 2011 (main data collection) across the three communities. Some key informant interviews were also conducted in Kutubdia Para in May 2013 to update some qualitative data.

Table 9.3 Characteristics of the three-study fishing communities

Characteristics	Padma	Kutubdia Para	Kutubdia Island	Source of data
Location	On mainland; in the western region of the coastal zone	On mainland; in the eastern region of the coastal zone	An island in the eastern region of the coastal zone	Pramanik (1983)
History of settlement	Most people have been living here for more than 100 years	People have migrated here from Kutubdia Island two decades ago	Most people have been living in this village for long time. Many of their neighbours migrated to Kutubdia Para	Reconnaissance study (updated during main data collection)
Land elevation (above the mean sea level)	0.9–2.1 m	Less than 1 m	Less than 0.5 m	Iftekhar and Islam (2004)
Proximity to town	8 km from a local town	6 km from a big tourist town	No town or tourist area nearby	Reconnaissance study (updated during main data collection)
Cyclone shelter	Two cyclone shelters present	No cyclone shelter present but many buildings present in nearby town	Few cyclone shelters present	Reconnaissance study (updated during main data collection)
Main livelihood activities	Fishing, fish processing, fish trading, shrimp post-larvae collecting, agriculture farming and labouring	Fish processing, fishing, fish trading, shrimp post-larvae collecting and labouring	Fish processing, fishing, salt producing, agriculture farming, fish trading, shrimp post-larvae collecting and labouring	Reconnaissance study (updated during main data collection)
Fishing area	Catch fish mainly from near-shore areas and "west of swatch of no ground"[a]	Catch fish mainly from near-shore areas and "south patches"[a]	Catch fish mainly from near-shore areas and "south patches"[a]	Reconnaissance study (updated during main data collection)
Cyclones and floods	Few major cyclones and floods in the past 40 years and some minor cyclones each	Few major cyclones and floods in the past 40 years but at different times than Padma and some minor	Few major cyclones and floods in the past 40 years at the same time like Kutubdia Para and some minor	Reconnaissance study (updated during main data collection)

(continued)

Table 9.3 (continued)

Characteristics	Padma	Kutubdia Para	Kutubdia Island	Source of data
	year in the fishing area[b]	cyclones each year in the fishing area[b]	cyclones each year in the fishing area[b]	
Trend of sea level rise	2.9 mm/year	1.4 mm/year	2.1 mm/year	CEGIS (2006)
Soil erosion	Medium	Low	Very high	Pramanik (1983)

[a]Fishing grounds in the Bay of Bengal
[b]Based on intensity, cyclones are different types (together also known as cyclonic disturbances). Major cyclones include severe cyclonic storm (48–63 knots), very severe cyclonic storm (64–119 knots) and super cyclonic storm ($\geq$ 120 knots), while minor cyclones include depression (17–27 knots), deep depression (28–33 knots) and cyclonic storm (34–47 knots) (WMO 2012)

Results and Discussion

While assessing vulnerability to climate variability and change, this research highlights that the level of livelihood vulnerability to climate variability and change differs not only between communities but also between different household groups within a community, depending on their level of exposure, sensitivity and adaptive capacity. Exposure to floods and cyclones; sensitivity (such as dependence on small-scale marine fisheries for livelihoods); and lack of adaptive capacity in terms of physical, natural and financial capital and diverse livelihood strategies construe livelihood vulnerability in different ways depending on the context. As a whole, Padma's households are more vulnerable than those of Kutubdia Para. This is because the households in Padma are much more exposed to climate shocks and stresses than their Kutubdia Para's counterparts. However, they are neither more sensitive nor have they less adaptive capacity than their Kutubdia Para's counterparts. Thus, the most exposed community is not necessarily the most sensitive or least able to adapt because livelihood vulnerability is a result of combined but unequal influences of bio-physical and socio-economic characteristics of communities and households. But within a fishing community, where households are similarly exposed, higher sensitivity and lower adaptive capacity combine to create higher vulnerability. As evident in this study, within Padma, where the several groups of households are similarly exposed to climate shocks and stresses, higher sensitivity and lower adaptive capacity combine to produce higher livelihood vulnerability. A similar trend is also found between the household groups in Kutubdia Para. For detailed results (including data) about vulnerability of fishery-based livelihoods to climatic impacts, see Islam et al. (2014a).

Assessing the outcomes of climate-induced internal migration, this research also demonstrates that migration *may* be a viable strategy to cope with or adapt to climate variability and change. The migrant households (in Kutubdia Para) are less exposed to climate shocks and stresses than their non-migrant counterparts (in Kutubdia Island). They also have more livelihood assets and have better access to

Table 9.4 Summary of key methods used to address the aim and each objective

Research aim and objectives	Methods for primary data collection	Methods for data analysis
Aim: to assess the vulnerability and adaptation of three Bangladeshi coastal small-scale fishing communities to the impacts of climate variability and change	Mixed methods	Quantitative and qualitative
Objective 1: to assess the vulnerability of fishery-based livelihoods to the impacts of climate variability and change in two coastal fishing communities (Padma and Kutubdia Para) and their households	Structured household questionnaires, oral history interviews, key informant interviews, vulnerability matrices and focus group discussions (FGDs)	Cluster analysis, construction of vulnerability indices by composite index approach, t-test, z-test and ANOVA for quantitative data, and content analysis (by coding) for qualitative data
Objective 2: to examine how climate-induced permanent migration has impacted vulnerability and adaptation of a coastal fishing community (Kutubdia Para) by comparing with the residual of its original community (Kutubdia Island)	Structured household questionnaires, oral history interviews, key informant interviews, vulnerability matrices and FGDs	Cluster analysis, descriptive statistics and z-test for quantitative data, and content analysis (by coding) for qualitative data
Objective 3: to identify and characterise limits and barriers to adaptation of fishing activities to cyclones and examine interactions between them in two fishing communities (Padma and Kutubdia Para)	Structured household questionnaires, oral history interviews, vulnerability matrices and FGDs	Descriptive statistics for quantitative data and content analysis (by coding) for qualitative data

them. They enjoy higher incomes, better health and better access to water supply, health and educational services, technology and markets than the households who did not migrate. Although they are comparatively more dependent on climate-sensitive marine fisheries, they are in a better position to divert away from climate-sensitive livelihoods through investing their higher incomes in other activities or in the building of human capital to enable this diversification. The non-migrants have not been able to reduce their vulnerability or to increase their ability to respond to climatic shocks and stresses. They face difficulty in continuing their livelihoods, and most of them are keen to migrate away from the island. However, they cannot do so due to lack of assets and outside support, and the uncertainty of livelihoods at the destination. They have become trapped in a more

Table 9.5 Sample sizes for data collection from the study communities

Communities	Structured household questionnaires	Oral history interviews	Vulnerability matrices	Key informant interviews	Focus group discussions (FGDs)
Padma	100	22	5	11 (5 during reconnaissance study + 6 during main data collection)	7 (2 during reconnaissance study + 5 during main data collection)
Kutubdia Para	100	21	4	14 (5 during reconnaissance study + 6 during main data collection + 3 later to update some data)	6 (2 during reconnaissance study + 4 during main data collection)
Kutubdia Island	50	17	3	6 (only during main data collection)	4 (only during main data collection)
Total	250	60	12	31	17

vulnerable position and are unable to exit this situation on their own. For detailed results about outcomes of climate-induced internal migration, see Islam et al. (2014b).

Examining the limits and barriers to adaptation of fishing activities to cyclones across the case study sites (Padma and Kutubdia Para), this study further illustrates that adaptation is constrained by multiple interacting limits and barriers. The limits include physical characteristics of climate and sea such as higher frequency and duration of cyclones, and hidden sandbars. Barriers include technologically poor boats, inaccurate weather forecasts, poor radio signals, lack of access to credit, low incomes, underestimation of cyclone occurrence, coercion of fishermen by the boat owners and captains, lack of education, skills and livelihood alternatives, unfavourable credit schemes, lack of enforcement of fishing regulations and maritime laws, and lack of access to fish markets. These local and wider scale factors interact in complex ways and constrain completion of fishing trips, coping with cyclones at sea, safe return of boats from sea, timely responses to cyclones and livelihood diversification. For detailed results about limits and barriers to adaptation of fishing communities, see Islam et al. (2014c).

Based on the findings of this study, it can be said that the coastal fishing communities have been impacted by several climatic shocks and stresses and they have traditionally coped with or adapted to the normal range of climate impacts but not always sufficiently well. Thus, autonomous adaptation is not sufficient for them to address the current climate variability and change. In the coming decades, the vulnerability of fishery-based livelihoods may substantially increase because of climate change. Almost all the livelihood assets and strategies of fishing communities will face the impacts of sea level rise, land erosion, cyclones and associated flooding, as these are predicted to be exacerbated due to climate change (2007b),

Met Office (2011). Without adaptation, increased levels of cyclones and floods will result in greater loss of life in the coastal areas and at sea, greater damage to fishing boats, gear and other household assets, especially a loss of fishery-related income. An accelerated sea level rise as projected during this century (MoEF 2005) will result in permanent inundation and accelerated erosion of the land base of Bangladeshi coastal communities. Alterations in temperature and rainfall can have direct impacts on the capacity for fish drying, which is the most common fish processing activity in this region. In short, future climate change is predicted to impact outside the normal range, for which additional adaptation will be needed for the fishing communities (FAO 2008; Allison et al. 2005).

Policy Implications

The findings of this study allow identification of a range of measures that could help address the impacts of current and future climate variability and change for the fishing communities in Bangladesh and potentially beyond. The findings are of particular relevance to the Government of Bangladesh's policy goal of "assess[ing] potential threats [of climate change] to the marine fish[eries] sector and develop [ing] adaptive measures".

Reduction of impacts, vulnerability or risks, increase in adaptive capacity or resilience, and facilitating adaptation actions and processes to climate variability and change for the fishing communities would require multifaceted measures. Global climate change mitigation is essential over the longer term to reduce exposure, overcome the limits to adaptation and build resilience, because adaptive capacity may be limited to only lower levels of climate change ($\leq 2–3$ °C) (Climate Change 2007a). Improved weather forecasting, warning and evacuation systems can reduce exposure to extreme events both in inland and at sea. Investment here would clearly impact on those that have no choice but to go to sea. In addition, investment in hard infrastructure such as concrete sea walls surrounding the communities could protect them from storm surge, sea level rise and land erosion. However, given Bangladesh's limited economic resources, investment in hard infrastructure is unlikely in the near future.

Reducing exposure needs to be complemented with reducing sensitivity, increasing adaptive capacity and supporting adaptation processes through planned adaptation. Ensuring improved livelihood outcomes of fishing communities by augmenting their livelihood assets and improving access to them, and helping to diversify livelihood strategies could be helpful in this regard. Modernisation of fishing technology (such as improved quality fishing boats) and radio signalling could not only help save lives in the sea but also reduce the damage to fishing assets particularly from cyclones. Modernisation of fish drying technology (such as more use of solar tent driers) will also be required in future to adapt the fish drying process to increased variations in temperature and rainfall. Access to less expensive financial credit through institutional reform could help transform fishing and fish drying technologies, build human capital, facilitate necessary migration, assist diversification of livelihoods and prevent maladaptation (by helping to build comparatively expensive robust fishing boats) in the fishing business. This institutional reform would include provision for flexible collateral, flexible credit repayment schemes and easing the application process so that less educated/uneducated people could apply for credit easily and they would not need to have special knowledge of the credit systems or pre-existing relationships with credit providers to get credit. Institutional reform is also required to improve enforcement of maritime laws (such as through increasing the capacity of law enforcing members and removing corruption among them) and access to fish markets (such as through reducing middlemen in the marketing chain and reducing the charge for selling fish in the auction market) to help reduce the overall costs of the fishing business. Enforcement of fishing regulations and provision of insurance would increase the safety of fishermen. Building human capital, such as through investment in education and skills development, could particularly help to diversify livelihoods which in turn would help individuals and associated households to become less reliant on climate-sensitive marine fisheries. The fishery systems could also be made less climate-sensitive by conserving fishery resources to ensure sustainable fish stocks and incomes for the future. Protection of near-shore fish stocks is a priority given the lack of fish reported in this area—this would also divert the offshore fishermen into near-shore areas where it is easier to respond to cyclones.

However, caution should be maintained as some adaptation strategies may exacerbate existing problems or may be maladaptive to other systems. For example, the construction of sea walls to protect the communities, proposed above, may change the offshore sediment balance and increase erosion in adjacent coastal areas (Eriksen et al. 2007). Black et al. (2011b) have warned that the condition of the people who are unable or unwilling to migrate to address climate impacts may be exacerbated by maladaptive policies designed to prevent migration. To ensure that existing problems are not exacerbated, Fazey et al. (2009, p. 414) suggest that adaptation must: address both human-induced and biophysical drivers of change, maintain a diversity of future response options and nurture the kinds of human capacities that enable the uptake of those response options. It is also necessary to ensure that adaptation does not lead to greater disparity and inequity between households or social groups, especially to achieve sustainable development. Thus, rather than adopting more radical options, mainstreaming of adaptation within a wider development arena is often preferred (e.g. Stringer et al. 2009; Brown 2011).

Scaling-up and Transferability of Findings

An important issue within vulnerability and adaptation research and within the development community is the transferability of adaptation responses and the scaling-up lessons learnt from local case studies. Development practitioners are calling for lessons to be learnt from case study research in order to help facilitate adaptation in other communities. There have been calls within the adaptation and climate change community for greater local-level case study and comparative research in order to identify generic aspects of adaptation, especially the key features leading to successful adaptation (Reid and Vogel 2006; Smit and Wandel 2006).

Smit and Wandel (2006) suggest that research on practical adaptation initiatives can be used to compare across communities and societies in order to identify aspects of adaptation that are effective. By carrying out local-level studies in three communities, it was possible to investigate whether there are generic findings that may have implications for other communities in this region. These types of findings can be commonly scaled up from the case study communities to their larger archetypal livelihood region (Iwasaki et al. 2009). Developing countries, in particular the coastal people of South and Southeast Asia whose primary income source is the fishery resources of the Bay of Bengal, can be considered as the larger archetypal livelihood region of this study. The Bay of Bengal fisheries cater for people from seven countries: Bangladesh, India, Malaysia, Maldives, Myanmar, Sri Lanka and Thailand. The climatic conditions of these countries are largely influenced by the Bay of Bengal. In a review, Townsley (2004) found that these fisheries-dependent people have some common characteristics such as high vulnerability to natural hazards such as cyclones and floods, exposure to seasonality of climate, prevalence of poverty, relatively high level of dependency on natural/ common property resources, use of traditional small-scale methods of fish exploitation and having strong market orientation to sell or exchange products for livelihoods. He also found some unique characteristics among some fishery-dependent people, such as the presence of ethnic groups and caste systems mostly on the northern and western side. Coulthard (2008) also found caste systems in south India which influenced adapting to environmental change in these fisheries. This caste system is not found in the current study communities. Therefore, the findings of this study may only *partly* be transferred and scaled up to other coastal fishing communities in the Bay of Bengal region as the characteristics of the population of this study only have some similarities with the characteristics mentioned in these countries.

However, Twyman et al. (2011) suggest that it is challenging to scale up the lessons from case study research to larger scales. When developing future institutional and local response interventions to facilitate adaptation to future climate change, Reid and Vogel (2006, p. 204) suggested "one size will not fit all". Thus, caution should be maintained about what lessons can be learned and transferred between case studies, especially as the impacts of climate change will be different between different places and social groups (Climate Change 2007a).

Conclusion

This research has assessed the vulnerability and adaptation of three coastal fishing communities to the impact of climate variability and change in Bangladesh. The case study approach has combined climate change vulnerability and adaptation approaches, composite index approach and SLA to develop the research framework and guide the overall data collection and analysis processes.

This study finds that the coastal fishing communities have been impacted by several climatic shocks and stresses and they have traditionally coped with or adapted to the normal range of climate impacts but not always sufficiently well. Thus, autonomous adaptation is not sufficient for them to address the current climate variability and change. Reduction of impacts, vulnerability or risks, increase in adaptive capacity or resilience, and facilitating adaptation actions and processes to climate variability and change for the fishing communities would require multifaceted measures. Global climate change mitigation is essential over the longer term to reduce exposure, overcome the limits to adaptation and build resilience, because adaptive capacity may be limited to only lower levels of climate change (≤ 2–3 °C). To reduce exposure, other measures need to be taken, such as improved weather forecasting, warning and evacuation systems, and building concrete wall surrounding the communities. Reducing exposure needs to be complemented with reducing sensitivity, increasing adaptive capacity and supporting adaptation processes through planned adaptation. The findings of this study allow identification of a range of measures that could help address the impacts of current and future climate variability and change for the fishing communities in Bangladesh and potentially beyond.

Acknowledgements This paper is part of a PhD study funded by the Commonwealth Scholarship Commission. The fieldwork was supported by the ESRC Centre for Climate Change Economics and Policy (CCCEP); Sustainability Research Institute of the University of Leeds; Carls Wallace Trust, UK; and Annesha Group, Bangladesh. The author is grateful to his supervisors for their guidance at all stages of the PhD.

References

Adger WN, Huq S, Brown K, Conway D, Hulme M (2003) Adaptation to climate change in the developing world. Prog Dev Stud 3(3):179–195
Adger WN, Agrawala S, Mirza MNQ, Conde C, O'Brien K, Pulhin J, Pulwarty R, Smit B, Takahashi K (2007) Assessment of adaptation practices, options, constraints and capacity. In: Parry ML, Canziani OF, Palutikof JP, van der Linden PJ, Hanson CE (eds) Climate change 2007: impacts, adaptation and vulnerability. Contribution of working group II to the fourth assessment report of the intergovernmental panel on climate change. Cambridge University Press, Cambridge, UK, pp 717–743
Adger WN, Lorenzoni I, O'Brien KL (2009) Adapting to climate change: thresholds, value, Governance. Cambridge University Press, Cambridge

Aghazadeh E (1994) Fisheries: socio-economic analysis and policy—assistance to Fisheries Research Institute, Bangladesh Project reports FAO, p 127

Agrawala S, Ota T, Ahmed AU, Smith J, van Aalst M (2003a) Development and climate change in Bangladesh: focus on coastal flooding and the Sundarbans. Organisation for Economic Co-operation and Development, Paris, France

Agrawala S, Raksakulthai V, van Aalst M, Larsen P, Smith J, Reynolds J (2003b) Development and climate change in Nepal: focus on water resources and hydropower. Organisation for Economic Co-operation and Development, Paris

Ahmed S (2002) Final report of the coastal and marine fisheries management improvement project (in Bengali), Department of Fisheries, Government of Bangladesh, Dhaka

Ahmed AU, Neelormi S (2008) Livelihoods of coastal fishermen in peril: in search of early evidence of climate change induced adverse effects in Bangladesh. Centre for Global Change, Dhaka

Ahmed MK, Ameen M, Sultana S (2002) Impact of global climate change and variability on fisheries resources of Bangladesh, water and Climate in Bangladesh. IUCN Bangladesh, Dhaka

Ahmed MS, Alam MS, Akther H (2009) Livelihood assessment of a jatka fishing community at north Srirumthi village, Cnandpur, Bangladesh. South Pac Stud 29(2):1–12

Ahmmed h (2007) Pre- and post-harvest handling and processing procedures for the production of safe and high quality fresh fisheries products in Bangladesh. Final Project 2007, United Nation University, Fisheries Training Program, Reykjavik, Iceland

Ahmed MK, Halim S, Sultana S (2012) Participation of women in aquaculture in three coastal districts of Bangladesh: approaches toward sustainable livelihood. World J Agricult Sci 8 (3):253–268

Akter N, Ahmed K, Islam MS, Koo S (2009) An economic study on small scale marine fishing in Teknaf of Cox's Bazar district of Bangladesh. South Asia Res 15(2):289–313

Ali MY (1999) Fish resources vulnerability and adaptation to climate change in Bangladesh. In: Huq S, Karim Z, Asaduzzaman M, Mahtab F (eds) Vulnerability and adaptation to climate change for Bangladesh. Kluwer Academic Publishers, Dordrecht, pp 113–124

Allison EH, Adger WN, Badjeck MC, Brown K, Conway D, Dulvy NK, Halls A, Perry A, Reynolds JD (2005) Effects of climate change on the sustainability of capture and enhancement fisheries important to the poor: analysis of the vulnerability and adaptability of fisherfolk living in poverty, Fisheries Management Science Programme. MRAG for Department for International Development, UK, London, p 167

Allison EH, Perry AL, Badjeck M-C, Adger WN, Brown K, Conway D, Halls AS, Pilling GM, Reynolds JD, Andrew NL, Dulvy NK (2009) Vulnerability of national economies to the impacts of climate change on fisheries. Fish Fish 10(2):173–196

Bala BK, Mondol MRA (2001) Experimental investigation on solar drying of fish using solar tunnel dryer. Drying Technol 19(2):427–436

Barnett J, O'Neill SJ (2012) Islands, resettlement and adaptation. Nat Clim Change 2(1):8–10

BCCSAP (2008) Bangladesh climate change strategy and action plan, Ministry of Environment and Forests, People's Republic of Bangladesh

Bengtsson L, Hodges KI, Esch M, Keenlyside N, Kornblueh L, Luo J-J, Yamagata T (2007) How may tropical cyclones change in a warmer climate? Tellus A 59(4):539–561

Black R, Adger WN, Arnell NW, Dercon S, Geddes A, Thomas D (2011a) The effect of environmental change on human migration. Glob Environ Change 21(Supplement 1(0)):S3–S11

Black R, Bennett SRG, Thomas SM, Beddington JR (2011b) Migration as adaptation. Nature 478 (7370):447–449

BOBP (1985) Marine small-scale fisheries of Bangladesh: a general description. Development of Small-Scale Fisheries, Bay of Bengal Programme Madras, India

Brander K (2006) Assessment of possible impacts of climate change on fisheries. WBGU, Berlin

Brown K (2011) Sustainable adaptation: an oxymoron? Clim Dev 3(1):21–31

CEGIS (2006) Impact of sea level rise on land use suitability and adaptation options. Final draft report. Submitted to the Ministry of Environment and Forest, Government of Bangladesh and

United Nations Development Programme, Centre for Environmental Geographic Information Services, Dhaka

Chowdhury IU (2009) Fishing communities in coastal Bangladesh: An overview of sustainable livelihoods. http://www.allacademic.com/meta/p_mla_apa_research_citation/0/2/3/4/2/p.23421_index.html. Accessed 16 Oct 2009

Chowdhury SR, Hossain MS, Shamsuddoha M, Khan SMMH (2012) Coastal fishers' livelihood in peril: sea surface temperature and tropical cyclones in Bangladesh. Center for Participatory Research and Development, Dhaka, p 54

IPCC, Climate Change (2007) Impacts, Adaptation And Vulnerability. Contribution of Working Group II to the fourth assessment report of the intergovernmental panel on climate change. In: Parry ML, Canziani OF, Palutikof JP, van der Linden PJ, Hanson CE (eds) Cambridge University Press, Cambridge

IPCC, Climate Change (2007) The physical science basis. Contribution of working group I to the fourth assessment report of the intergovernmental panel on climate change. In: Solomon S, Qin D, Manning M, Chen Z, Marquis M, Averyt KB, Tignor M, Miller HL (eds) Cambridge University Press, Cambridge

Coulthard S (2008) Adapting to environmental change in artisanal fisheries—insights from a South Indian Lagoon. Glob Environ Change 18(3):479–489

Coulthard S (2009) Adaptation and conflict within fisheries: insights for livling with climate change. In: Adger WN, Lorenzoni I, O'Brien KL (eds) Adapting to climate change: thresholds values, governance. Cambridge University Press, Cambridge

Dasgupta S, Laplante B, Murray S, Wheeler D (2011) Exposure of developing countries to sea-level rise and storm surges. Clim Change 106(4):567–579

DoF (2009) Fisheries statistical yearbook of Bangladesh (2007-2008), Fisheries Resources Survey System, Department of Fisheries, Government of Bangladesh, Dhaka, p 49

DoF National (2012) fisheries week. Department of Fisheries, Government of Bangladesh, Dhaka

Eisenhardt KM (1989) Building theories from case study research. Acad Manage Rev 14(4):532–550

Emanuel KA (1987) The dependence of hurricane intensity on climate. Nature 326(6112):483–485

Emanuel K, Sundararajan R, Williams J (2008) Hurricanes and global warming: Results from downscaling IPCC AR4 simulations. Bull Am Meteor Soc 89(3):347–367

Eriksen SEH, Klein RJT, Ulsrud K, Naess LO, O'Brien K (2007) Climate change adaptation and poverty reduction: key interactions and critical measures. Report prepared for the Norwegian Agency for Development Cooperation (Norad), GECHS Report 2007:1, Global Environmental Change and Human Security, Oslo

Fankhauser S (1996) The potential costs of climate change adaptation. In: Smith J, Bhatti, Menzhulin R, Benioff MI, Budyko M, Campos BJ, Rijsberman F (eds) Adapting to climate change: an international perspective. Springer, New York, USA, pp 80–96

FAO (2008) Report of the FAO expert workshop on climate change implications for fisheries and aquaculture. Food and Agriculture Organisation, Rome

FAO (2012) The state of world fisheries and aquaculture 2012, Fisheries and Aquaculture Department, Food and Agriculture Organization of the United Nations, Rome

Fazey I, Gamarra JGP, Fischer J, Reed MS, Stringer LC, Christie M (2009) Adaptation strategies for reducing vulnerability to future environmental change. Front Ecol Environ 8(8):414–422

Field CB, Barros VR, Mach K, Mastrandrea M (2014) Climate change 2014: impacts, adaptation, and vulnerability. Cambridge University Press Cambridge, New York

GOS (2011) Foresight: migration and global environmental change. Final project report, The Government Office for Science, UK, London

Gray WM (1968) Global view of the origin of tropical disturbances and storms. Mon Weather Rev 96(10):669–700

Han M, Zhao MH, Li DG, Cao XY (1999) Relationship between ancient channel and seawater intrusion in the south coastal plain of the Laizhou Bay. J Nat Disasters 8:73–80

Hasan M, Billah MM, Roy TK (2004) Tourism and fishing community of Kuakata: a remote coastal area of Bangladesh, Part—1, Support for University Fisheries Education and Research Project, DFID

Iftekhar MS, Islam MR (2004) Managing mangroves in Bangladesh: a strategy analysis. J Coast Conserv 10:139–146

IPCC (2012) Managing the risks of extreme events and disasters to advance climate change adaptation. A special report of working groups I and II of the intergovernmental panel on climate change. In: Field CB, Barros V, Stocker TF, Qin D, Dokken DJ, Ebi KL, Mastrandrea MD, Mach KJ, Plattner G-K, Allen SK, Tignor M, Midgley PM (eds) Cambridge University Press Cambridge

Islam D, Mustafa MG, Wahed MA, Khaleque MA, Naser MN, Islam MM (2007) Comparative assessment of traditional sun dried and solar tunnel dried pomfret (*Pampus argenteus*) under different storage conditions, Bangladesh. J Zool 35(2):331–340

Islam D, Mustafa MG, Wahed MA, Khaleque MA, Islam MM (2008) Investigation on the performance of solar tunnel dryer for dehydration of *Gudusia chapra* and *Channa striatus* over traditional sun drying. Dhaka Univ J Biol Sci 17(2):169–172

Islam MM, Sallu S, Hubacek K, Paavola J (2014a) Vulnerability of fishery-based livelihoods to the impacts of climate variability and change: insights from coastal Bangladesh. Reg Environ Change 14(1):281–294

Islam MM, Sallu S, Hubacek K, Paavola J (2014b) Migrating to tackle climate variability and change? Insights from coastal fishing communities in Bangladesh. Clim Change 124(4):733–746

Islam M, Sallu S, Hubacek K, Paavola J (2014c) Limits and barriers to adaptation to climate variability and change in Bangladeshi coastal fishing communities. Mar Policy 43:208–216

Iwasaki S, Razafindrabe BHN, Shaw R (2009) Fishery livelihoods and adaptation to climate change: a case study of Chilika lagoon, India. Mitig Adapt Strat Glob Change 14(4):339–355

Jones L, Boyd E (2011) Exploring social barriers to adaptation: insights from western Nepal. Glob Environ Change 21(4):1262–1274

Karim MF, Mimura N (2008) Impacts of climate change and sea-level rise on cyclonic storm surge floods in Bangladesh. Glob Environ Change 18(3):490–500

Khan TMA, Singh OP, Rahman MDS (2000) Recent sea level and sea surface temperature trends along the Bangladesh coast in relation to the frequency of intense cyclones. Mar Geodesy 23:103–116

Libreo AM (1987) Women's roles in institutions and credit. In: Nash CE, Engle CR, Crosetti D (eds) Women in aquaculture - ADCP/REP/87/28. Proceedings of the ADCP/NORAD Workshop on Women in Aquaculture. Rome, Food and Agriculture Organisation

McDonald RE, Bleaken DG, Cresswell DR, Pope VD, Senior CA (2005) Tropical storms: Representation and diagnosis in climate models and the impacts of climate change. Clim Dyn 25(1):19–36

McLeman RA (2011) Settlement abandonment in the context of global environmental change. Glob Environ Change 21(Supplement 1(0)):S108–S120

McLeman R, Smit B (2006) Migration as an adaptation to climate change. Clim Change 76(1):31–53

Met Office (2011) Climate: observations, projections and impacts—Bangladesh, Met Office Hadley Centre and Department of Energy and Climate Change of the United Kingdom

Mirza MQ (2002) Global warming and changes in the probability of occurrence of floods in Bangladesh and implications. Glob Environ Change 12:127–138

Mirza MQ (2003) Three recent extreme floods in Bangladesh: a hydro-meteorological analysis. Nat Hazards 28(1):35–64

Mirza MQ, Dixit A (1997) Climate change and water management in the GBM Basins. Water Nepal 5:71–100

MoEF National (2005) adaptation programme of action (NAPA). Ministry of Environment and Forest, People's Republic of Bangladesh, Dhaka

MoEF (2009) National adaptation programme of action, Updated Version of 2005, Ministry of Environment and Forests, People's Republic of Bangladesh, Dhaka

MoEF (2012) Rio + 20: National report on sustainable development, Ministry of Environment and Forests, Peoples' Republic of Bangladesh, Dhaka

Morgan CL (2011) Limits to adaptation: a review of limitation relevant to the project "building resilience to climate change—coastal southeast Asia". IUCN, Gland, Switzerland

Mortreux C, Barnett J (2009) Climate change, migration and adaptation in Funafuti, Tuvalu. Glob Environ Change 19(1):105–112

Moser SC, Ekstrom JA (2010) A framework to diagnose barriers to climate change adaptation. Proc Natl Acad Sci 107(51):22026–22031

Myers N (2002) Environmental refugees: a growing phenomenon of the 21st century. Philos Trans R Soc London Ser B: Biol Sci 357(1420):609–613

Nicholls RJ, Marinova N, Lowe JA, Brown S, Vellinga P, de Gusmão D, Hinkel J, Tol RSJ (2011) Sea-level rise and its possible impacts given a 'beyond 4°C world' in the twenty-first century. Philos Trans R Soc A: Math Phys Eng Sci 369(1934):161–181

Nightingale A (2003) A feminist in the forest: situated knowledges and mixing methods in natural resource management. ACME 2(1):77–90

Oouchi K, Yoshimura J, Yoshimura H, Mizuta R, Kusunoki S, Noda A (2006) Tropical cyclone climatology in a global-warming climate as simulated in a 20 km-mesh global atmospheric model: frequency and wind intensity analyses. J Meteorol Soc Jpn Ser II 84(2):259–276

Paavola J (2008) Livelihoods, vulnerability and adaptation to climate change in Morogoro, Tanzania. Environ Sci Policy 11(7):642–654

Perry RI, Ommer RE, Barange M, Werner F (2010) The challenge of adapting marine social-ecological systems to the additional stress of climate change. Curr Opin Environ Sustain 2(5–6):356–363

Pielke R, Prins G, Rayner S, Sarewitz D (2007) Climate change 2007: lifting the taboo on adaptation. Nature 445(7128):597–598

Piguet E, Pécoud A, de Guchteneire P (2011) Migration and climate change: an overview. Refugee Surv Q 30(3):1–23

Pramanik MAH (1983) Remote sensing applications to coastal morphological investigations in Bangladesh. Jahangirnagar University, Dhaka, Bangladesh

Pramanik MAH (1988) Methodology and techniques of studying coastal systems: SPARRSO case studies, National Development Management Seminar held between, 3–4 Oct Dhaka

Rahman AKA (1994) The smallscale marine fisheries of Bangladesh, Department of Fisheries, Government of Bangladesh Dhaka

Rahman AA, Alam M, Alam SS, Uzzaman MR, Rashid M, Rabbani G (2007) Risks, vulnerability and adaptation in Bangladesh. Human Development Report 2007/2008. Fighting climate change: Human solidarity in a divided world, UNDP

Rashid HE (1991) Geography of Bangladesh. The University Press Limited Dhaka

Reid P, Vogel C (2006) Living and responding to multiple stressors in South Africa—glimpses from KwaZulu-Natal. Glob Environ Change 16(2):195–206

Shahid S (2010) Recent trends in the climate of Bangladesh. Clim Res 42(3):185–193

Smit B, Wandel J (2006) Adaptation, adaptive capacity and vulnerability. Glob Environ Change 16(3):282–292

Smith K (1996) Environmental hazards: assessing risk and reducing disaster. Routledge, London, United Kingdom

Stern N (2007) The economics of climate change: the Stern review. Cambridge University Press, Cambridge

Stocker TF (2014) Climate change 2013: the physical science basis: Working Group I contribution to the Fifth assessment report of the Intergovernmental Panel on Climate Change, Cambridge University Press

Stringer LC, Dyer JC, Reed MS, Dougill AJ, Twyman C, Mkwambisi D (2009) Adaptations to climate change, drought and desertification: local insights to enhance policy in southern Africa. Environ Sci Policy 12(7):748–765

Sugi M, Murakami H, Yoshimura J (2009) A reduction in global tropical cyclone frequency due to global warming. SOLA 5:164–167

Sultana P, Thompson P (2008) Gender and local floodplain management institutions: a case study from Bangladesh. J Int Dev: J Develop Stud Assoc 20(1):53–68

Townsley P (2004) Review of coastal and marine livelihoods and food security in the Bay of Bengal large marine ecosystem region, Bay of Bengal Large Marine Ecosystem Programme. Bay of Bengal Programme, Phuket, Thailand

Twyman C, Fraser EDG, Stringer LC, Quinn C, Dougill AJ, Ravera F, Crane TA, Sallu SM (2011) Climate science, development practice, and policy interactions in dryland agroecological systems. Ecol Soc 16(3):14

Vecchi GA, Soden BJ (2007) Effect of remote sea surface temperature change on tropical cyclone potential intensity. Nature 450(7172):1066–1070

Warrick RA, Oerlemans J (1990) Sea level rise. In: Houghton JL, Jenkins GJ, Ephraums JJ (eds) Climate change: the IPCC scientific assessment. Cambridge University Press, Cambridge, pp 257–281

Westlund L, Poulain F, Bage H, van Anrooy R (2007) Disaster response and risk management in the fisheries sector. Food and Agriculture Organization of the United Nations, Rome

WMO (2012) Coordination: review of terminology/classification of tropical cyclones, The seventh tropical cyclone RSMCS/TCWCS Technical coordination meeting, World Meteorological Organization, West Java, Indonesia

World Bank (2013a) Turn down the heat: climate extremes, regional impacts, and the case for resilience. A report for the World Bank by the Potsdam Institute for Climate Impact Research and Climate Analytics, World Bank, Washington, DC

World Bank (2013b) World development indicators: Bangladesh. http://data.worldbank.org/country/bangladesh#cp_wdi. Accessed 04 July 2013

Yin RK (1984) Case study research: design and methods. Sage Publications, Beverly Hills, CA

Yu J, Wang Y, Hamilton K (2010a) Response of tropical cyclone potential intensity to a global warming scenario in the IPCC AR4 CGCMs. J Clim 23(6):1354–1373

Yu W, Alam M, Hassan A, Khan AS, Ruane A, Rosenzweig C, Major D, Thurlow J (2010) Climate change risks and food security in Bangladesh, Earthscan Londo

Zhao M, Held IM, Lin S-J, Vecchi GA (2009) Simulations of global hurricane climatology, interannual variability, and response to global warming using a 50-km resolution GCM. J Clim 22(24):6653–6678

Chapter 10
Climate Change and Lightning Risk in Bangladesh

Fahmida Kabir and Md. Jakariya

Abstract In developing countries, lightning hazard is an underrated natural hazard despite having the potential to cause major loss and damages to human life and property and Bangladesh is not an exception. The existing national database of Bangladesh lacks information on lightning casualties. Hence, five years of database on lightning-related deaths and injuries from 2011 to 2016 was constructed through an innovative data mining process. An average of 913 casualties was identified, with an average of 182 people being affected by lightning occurrences each year in Bangladesh. The largest death toll was found among the male population (74%) compared to the females (26%), as males are more involved with labor-intensive agricultural practices in a developing country like Bangladesh. Most casualties occurred during the pre-monsoon (March–May) and monsoon (June–September) seasons with lightning incidents occurring mostly between morning (0600 LST) and afternoon (1800 LST). The most vulnerable age groups were found to be from 16 to 30 and 31 to 45 followed by <16, 46–60 and >60. Outdoor activities accounted for the highest number of lightening causalities; activities mostly involved agricultural practices followed by open area activities. Indoor dwellings also had significant amount of casualty especially in the veranda/balcony and while sleeping. The spatial distributions of lightning casualties were determined by GIS mapping; districts with no, low, moderate and high casualties were determined. Northwestern (Chapainawabganj) and northeastern districts (Kishoreganj and Moulavibazar) of Bangladesh accounted for the highest number of casualties. This study will therefore provide useful information in developing lightning safety measures in Bangladesh.

Keywords Lightning · Lightning casualties · Spatial distribution

F. Kabir · Md. Jakariya (✉)
Department of Environmental Science and Management, North South University,
Bashundhara, Dhaka, Bangladesh
e-mail: md.jakariya@northsouth.edu

© Springer Nature Switzerland AG 2021
Md. Jakariya and Md. N. Islam (eds.), *Climate Change in Bangladesh*,
Springer Climate, https://doi.org/10.1007/978-3-030-75825-7_10

Introduction

Lightning is one of the most powerful atmospheric phenomena that have fascinated mankind throughout the history. It has the potential to cause serious life-threatening injuries (Raga et al. 2014) and economic damages (Illiyas et al. 2014) in the society; therefore, it is important for people and administration to recognize lightning as a serious threat to the well-being of the human life and property. Globally, estimated 24,000 fatalities and 240,000 injuries occur due to lightning (Raga et al. 2014; Holle 2017). In Bangladesh, there has not been any natural disaster that goes unnoticed. However, major calamities can make some natural hazard go unnoticed (Sabur 2012), hence making us vulnerable to natural hazard such as lightning. According to Raga et al. (2014), minor- or moderate-scale disasters are not recognized as disaster due to their slow impact but this can build up overtime and turn out to have significant impact. The challenge with lightning is that it originates within the clouds, thus making it impossible for human to prevent it (Allaby 2003).

Lightning causes social and economic damages in the society with increased loss due to misleading information and lack of consciousness regarding lightning safety measures (Kithil 1995). Besides death, the most common injuries due to lightning strike involve an individual to suffer from burns, hearing problems, headaches, etc. (Gomes et al. 2006). Economic damage is another aspect of lightning. According to Uman (2008), in the USA 30% of its electric collapses because of lightning and it further costs billions of dollars of claims from the insurance companies due to property damage. In India, damages from the agricultural, housing and industrial sectors cost billions of rupees (Illiyas et al. 2014). The economic loss that occurs due to lightning has not been measured in Bangladesh, but as the country is developing it should soon start counting the economic loss (Gomes and Kithil n.d.).

Several factors contribute when it comes to an individual's exposure to lighting depending on their living conditions (houses that are not properly grounded and lack safety measures), working environment, occupational activities, awareness level, specific time of the year, knowledge of safety, etc. Literacy rate has been considered as a factor that can influence the amount of lightning accidents in a country. Countries with low literacy rate (e.g., India 66%, Pakistan 54% and Bangladesh 54%) had shown the trend of lack of compliance for the lightning protection regulations (Gomes and Kadir 2011). Low literacy rate contributes to lack of consciousness to follow the lighting safety directions, lack of willingness to follow the rules even if known and misinterpretation of lightning event. In Sri Lanka despite 80% of literacy rate in most of the districts, the deaths and injuries due to lightning are caused because of the reluctant nature of the people; the issue has not been as much as serious for them to consider it and take safety measures (Gomes et al. 2006).

Most common circumstances of lightning fatalities are recorded indoors as houses those were accounted for deaths and injuries due to lightning were not lightning-protected structures (Cardoso et al. 2014). According to Dlamini (2009) and Navarrete-Aldana et al. (2014), inappropriate construction, poorly earthed structures and lack of safe shelters and houses are responsible for casualties indoors. Lack of conscientiousness of professionals to follow the building code and to install safety measures is also another reason that causes indoor lightning incidents (Gomes and Kithil n.d.). Casualty from lighting greatly depends on the kind of activity an individual is involved with; countries whose socioeconomic activities greatly depend on the outdoor agricultural fields increase its people exposure to lightning hazard (especially during the early rainy season which requires a lot of activity in the growing lands) (Raga et al. 2014), because of which the population itself is being exposed to natural threat. In Brazil, most of the lighting casualties are related to agricultural activities comparing to that of the USA where most of the casualties are related to leisure or sports purposes such as golfing, fishing and camping. Activities differ by region to region in a given country, and one region may not have the same kind of lighting-related fatalities comparing to the other; e.g., in Brazil, fatalities related to telephone usage are more in the central region of the country than the northern region that has reported greater number of deaths related to playing football (Cardoso et al. 2014).

It is inevitable to assume lightning-caused casualties are highest among the populous regions; i.e., areas with high density of population will have the highest amount of people likely to be stricken by lightning, which has been true for many cases such as Turkey, Colombia and San Paulo (Brazil). As per Tilev-Tanriover et al. (2015), the number of lightning incidents was high for regions with highest density of population (Western Turkey) and lowest for the lowest population density (Central Turkey). In Colombia even though death ratio is larger for rural areas dispersedly located, it is said that the largest amount of death in Colombia due to lightning occurs in the urban regions as the population is concentrated toward those regions (Navarrete-Aldana et al. 2014); Sao Paulo of Brazil accounted for the highest number of deaths by lightning which the author directly linked to the largest amount of population of the state (Cardoso et al. 2014); but according to Raga et al. (2014), the author clearly suggests increased fatalities are not related to population density of an area, it is rather related to the exposure of an individual to thunderstorms, lack of education and working condition that makes an individual vulnerable to be stricken by lightning.

Lighting activity increases depending on different times of the year, and it is different for countries falling under different time zones; but in most of the cases lighting incidents occurred during summer. In the USA, lightening incidents reaches to its peak during summer having maximum during the month of July and decline just after the end of same month (Curran et al. 2000); **% of the fatalities in Swaziland occur during the summer season starting from October through February with a peak in November. However, there was no fatalities were recorded in Swaziland during the period of 2000–2009 (Dlamini 2009). In Mexico, causality incidents occur mostly in the first half of the rainy season (July and August) (Raga et al. 2014). Lightning incidents peak during the late spring (April–September) for Turkey having the highest incidents occurring during the months May and June (Tilev-Tanriover et al. 2015). In India, monsoon season

(June–September) accounts for 57% of fatalities with the incidents going down during winter (December–February) (Singh and Singh 2015).

In developed nations, lightning-related casualties have gone down due to the shifting of the population from rural to urban areas which does not require them to work on the labor-intensive agricultural activities (Holle 2016b). Developed nations are also economically advanced which allows them to be able to construct buildings/structures equipped with lightning safety systems, well-grounded houses and transports that are metal topped and fully enclosed (Dewan et al. 2017). Moreover, developed countries have been involved in numerous studies (e.g., the USA) and are quantifying lightning casualties that is enabling them to come up with better explanation on specific case studies, and providing better medical care to the injured and allowing them to put forward improved information on the phenomenon (Walsh et al. 2000).

In developing countries, lightning hazard is an underrated natural hazard despite having the potential to cause major loss and damages to human life and property (Dlamini 2009). As per Holle (2016a), a large number of people from the developing nations are involved with labor-intensive agricultural practices; they also reside in houses that are not well grounded or equipped with lightning safety devices that cause greater number of casualties in these nations (Dewan et al. 2017).

Review of the existing literature suggests that Asian developing countries have not received enough attention in addressing lightning-related phenomenon (Table 10.1). There has not been much studies in Bangladesh regarding lightning-related casualties except for few newspaper articles, and therefore, there is a lack of researches concerning lighting risk in Bangladesh, yet to be explored by its scientific community.

This research is an in-depth study of lighting in Bangladesh. The objective of the study is to find out variety of feature related to lightning-related deaths and injuries in Bangladesh bringing early and unwanted deaths to individuals considering factors such as gender, age groups, months, seasons, time, professions, activities and regions during the occurrence of the casualties.

Approach of the Study

The study is based on secondary qualitative information collected from reports published by the two most widely read online newspapers. Since the research topic was one of its kind and not many researches had been done, so very limited source of national reports was available. The only means to collect information was media-based literature from which data could be analyzed. The ultimate aim was to present the trend of increased casualties due to lightning and its physical attributes in Bangladesh in the last five years (2011–2016).

Secondary data source is an integral part of this study. Information was also collected from other published secondary sources for this study. Researcher who also conducted similar studies mostly depended on newspaper archives, online

Table 10.1 List of estimated lightning deaths and injuries identified and published by some of the authors carrying out similar study around the world

Location	Time frame	Casualties (deaths and injuries)	References
Australia	1824–1992	650 deaths identified	Coates et al. (1993)
Brazil	2000–2009	1321 deaths identified	Cardoso et al. (2014)
Canada		999 deaths (1921–2003) 47 fire-ignited deaths by lightning (1986–2001) Injuries (1986–2005)	Mills et al. (2008)
Colombia	2000–2009	757 deaths identified	Navarrete-Aldana et al. (2014)
India	2001–2014	31281 deaths identified	Selvi and Rajapandian (2016)
India	1979–2011	5259 deaths identified	Singh and Singh (2015)
Malawi (Nkhata Bay District)	2007–2010	11 deaths 44 injuries identified	Salerno et al. (2012)
Mexico	1979–2011	7300 deaths identified	Raga et al. (2014)
Singapore	1956–1979	80 deaths identified	Chao et al. (1981)
Swaziland	2000–2007	123 deaths identified	Dlamini (2009)
Turkey	1930–2014	895 deaths 149 serious injuries 535 injuries identified	Tilev-Tanriover et al. (2015)
UK	1993–1999	22 deaths 341 injuries	Elsom (2001)
USA	1959–1994	3239 deaths 9818 injuries	Curran et al. (2000)

newspaper data and death certificates from the government health database. Many on the other hand relied on the database available by the Ministry of Health (Cardoso et al. 2014; Chao et al. 1981) and Ministry of Home Affairs by accessing the information technology division (Selvi and Rajapandian 2016), weather databases and meteorological services (Tilev-Tanriover et al. 2015; Singh and Singh 2015; Chao et al. 1981). Information has also been accessed from organization such as Tornado and Storm Research Organisation (TORRO) by UK (Elsom 2001) and Storm Data by the USA (Curran et al. 2000).

Five years of study period was fixed from the year 2011 to 2016 from source_A and source_B. These two news sources were selected because of more data availability of the last five consecutive years compared to other online newspapers. Five years of study period was fixed due to limited research study period.

In-depth search of newspaper reports was studied, surveyed and noted in Excel sheets separately for source_A and source_B. This included the listing of number of months, deaths, injuries, gender casualties, age group, activity involved, weather events, duration, months, years, locations and analysis of regions having the most number of casualties related to lightning.

The data in Excel sheets were quantified through tally, and their values were determined. Those values were then used to make graphical charts, and the required results were found and attached. For mapping, Arc Geographical Information Software (GIS) was used to export region-wise casualty map for Bangladesh.

Results and Discussion

Lightning Casualties

Figure 10.1a, b shows five years of casualties (deaths and injuries) reported in source_A and source_B. From the figures, we can see that the casualties in the year 2011 were higher compared to the following years and then an increased casualty rate for the year 2016. The number of lightning deaths and injuries varies from year to year as seen in the figures. These numbers are estimated values since not all the casualties are reported due to limited geographical coverage; through Fig. 10.1a, b, it can be seen that deaths from lightning per year are above 25 to as much as 170 reported in a year for the last five years (2011–2016).

Gender Casualties

Figure 10.2a, b shows a visible difference in the gender variation in lightning casualties. It can be seen that comparatively more males are killed or affected by lightning than the females. Both the newspaper data analyses during the study period have found that the males accounted for 74% of lightning-related casualties (deaths and injuries) compared to females (that accounted for 26% of the casualties related to lightning) in Bangladesh.

A review on casualty due to lightning across the globe has been summarized in Table 10.2. Studies show that among the fatalities and injuries relating to lightning, the casualty percentages of males are greater than that of females. Most of the outdoor works are done by males for which males are more likely to be struck by

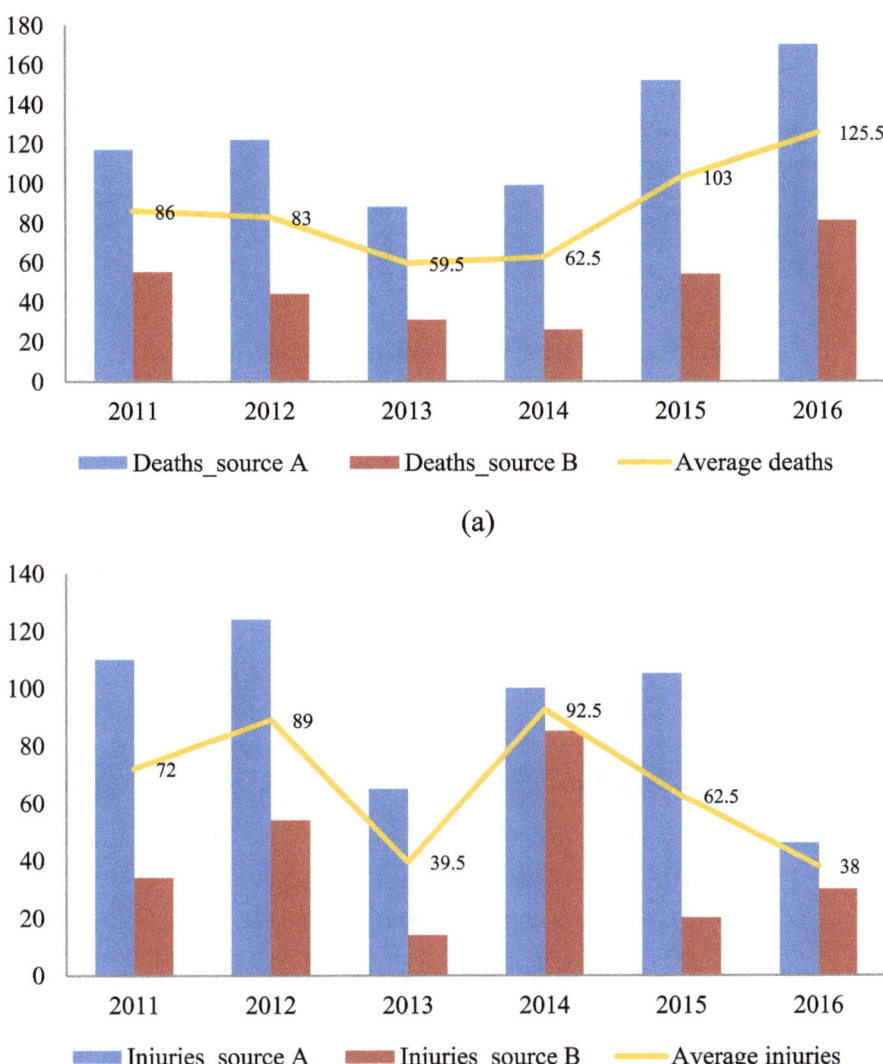

Fig. 10.1 **a** Deaths and **b** injuries reported from 2011 to 2016 through media analysis

lightning (Dlamini 2009; Tilev-Tanriover et al. 2015). Whereas in Turkey, probability of the females members to be striken by lightning is higher compare to other countries; as Turkey has 37% of females that contribute to the agricultural activities whereas for USA (2010) the percentage is 1%, Canada (2008): 1%, UK (2012) 1% and Mexico (2012): 4% (Tilev-Tanriover et al. 2015).

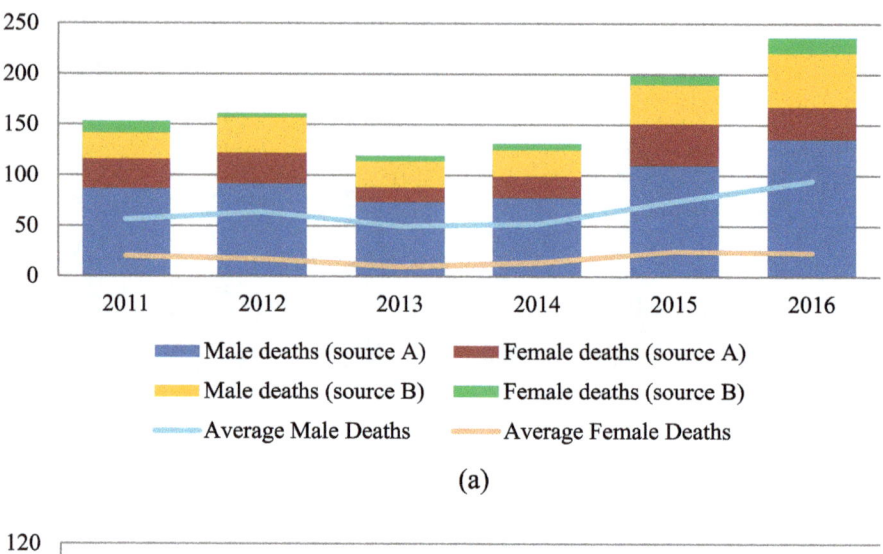

(a)

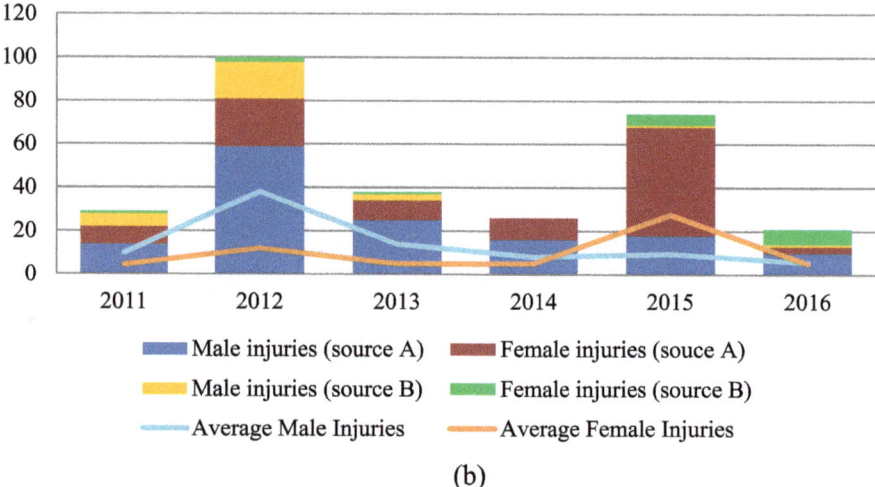

(b)

Fig. 10.2 a and **b** Gender casualties

Vulnerable Age Groups Involved in Lightning Incidents

People of different age groups that have been affected by lightning are represented in Fig. 10.3. Newspaper reports often lack on mentioning the age of people; hence, age groups have been set into five different categories with a difference of fifteen years of gap in between, starting with <16, 16–30, 31–45, 46–60 and >60. It is visible from the figures that the age group that is mostly affected by lightning is from 16 to 30 and 31 to 45 followed by <16, and 46–60 and the least affected are of >60 of age.

Table 10.2 Casualty percentage for male and female identified by different authors around the world

Location	Male (casualty %)	Female (casualty %)	Time frame	Authors
India	89	–	1979–2011	Singh and Singh (2015)
Colombia	80.3	19.7	2000–2009	Navarrete-Aldana et al. (2014)
Canada	84	16	1921–2003 (excluding the years from 1950 to 1964)	Mills et al. (2008)
UK	65	35	1993–1999	Elsom (2001)
Brazil	81	19	2000–2009	Cardoso et al. (2014)
Swaziland	68.3	–	2000–2007	Dlamini (2009)
USA	83	–	1950–1994	Curran et al. (2000)
Turkey	67	33	1930–2014	Tilev-Tanriover et al. (2015)

Fig. 10.3 Affected age groups due to lightning

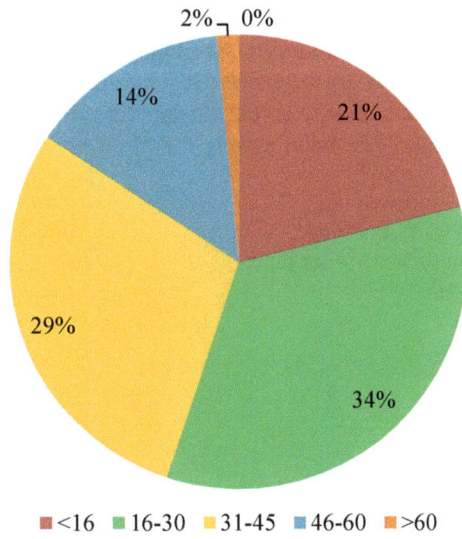

■ <16　■ 16-30　■ 31-45　■ 46-60　■ >60

Diurnal, Local Standard Time and Monthly Variation of Lightning Incidents

Most of the lightning incidents reported occurred during the morning and afternoon period followed by noon, evening and night and few percentage of incidents occurring during midnight and early hours of the day as shown in Fig. 10.4.

The local standard time for Bangladesh is +6GMT. The time has been recorded to show which hour has the most amounts of lightning incidents occurring. It can be seen from Fig. 10.5 that lightning-related casualties occurred mostly during early morning (6:00 a.m., 7:00 a.m., 8:00 a.m.) and late morning (10:00 a.m.) reporting peaked during the noon (12:00 p.m.) followed by afternoon and late hours of the day.

Monthly distributions of lightning incidents have been analyzed from the data sources. Incidents occurred mostly during pre-monsoon season (March–May) and monsoon (June–September) as shown in Fig. 10.6, reporting peaked during the month of April, May and June for each given year.

People from Different Occupations Involved in Lightning Casualties

Figure 10.7 represents a major portion of people from the farming profession who are likely to be affected by lightning followed by people from other professions, students, housewives and laborers. Fishermen on the other hand represent a very small percentage among the affected group of people.

Fig. 10.4 Diurnal distribution of lightning incidents

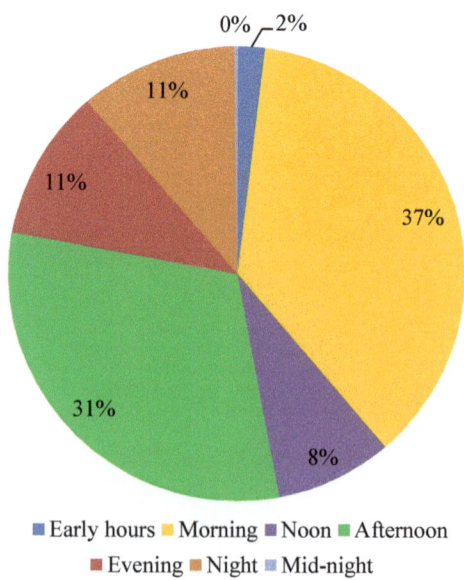

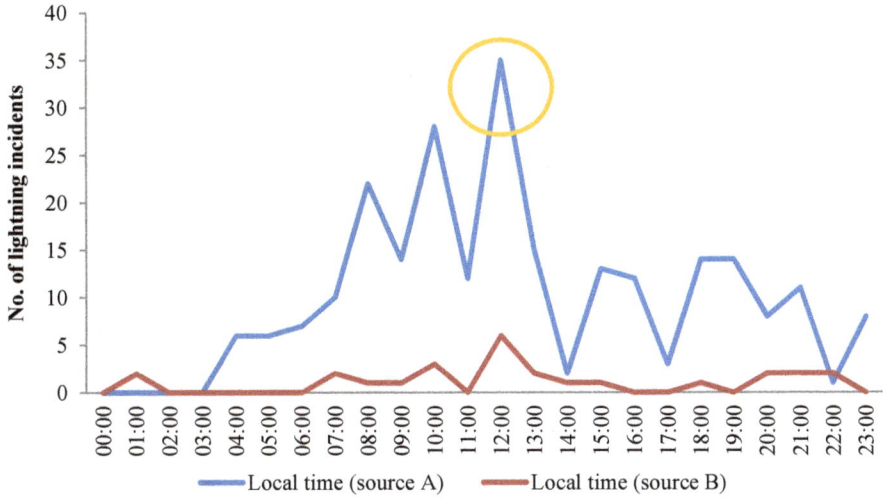

Fig. 10.5 Local standard time (LST) of lightning incidents

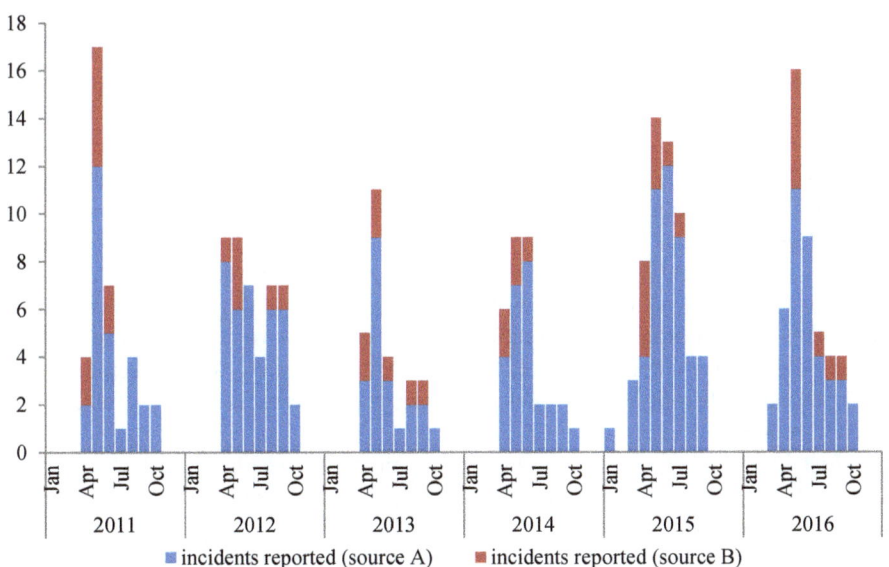

Fig. 10.6 Monthly variation of lightning incidents

Activities and Weather Events During Lightning Incidents

Majority of the lightning incidents occurred outdoor compared to indoors. Outdoor lightning activities have been divided into three categories: agricultural activities,

Fig. 10.7 Different
professions involved in
lightning casualties

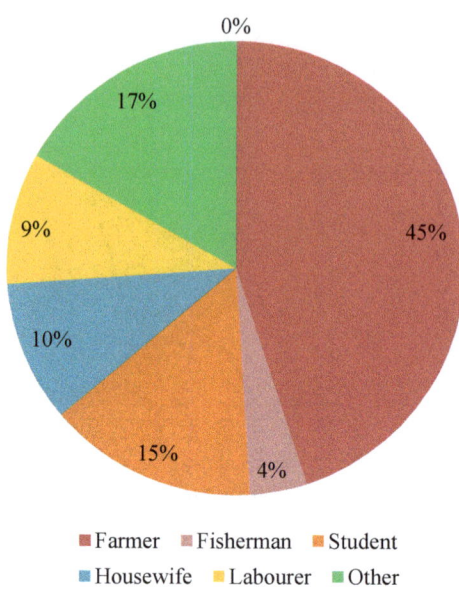

water-related activities and activities occurring in open area/space as shown in
Fig. 10.8a. It can be seen that majority of the lightning incidents are related to the
outdoor agricultural activities, e.g., harvesting/working/plowing and irrigating
paddy/farmlands/fields, followed by activities involving cattle herding or grazing
and mango collecting/plucking. Water-related activities mostly involved fishing and
traveling by the mean of boat. Open area activities involved going/returning from
market/school/bazaar followed by reactional activity such as playing cricket/
football in the open areas. Indoor activities are represented in Fig. 10.8b; even
though indoor activities account for less amount of lightning-related casualties it
still plays an important consideration. Figure 10.8b shows that majority of the
indoor lightning incidents occurred in the veranda/balcony, while sleeping inside
the house and when lightning struck on thatched and tin-roofed houses followed by
kitchen-related activities.

Most incidents reported lacked proper mention of weather events; among the
mentioned weather event, lightning incidents occurred mostly during the nor'-
wester/thunderstorm followed by heavy rainfall as shown in Fig. 10.9; we can also
see that the weather events had a certain percentage of drizzle and stormy winds
during the occurrence of lightning incidents.

Spatial Location of Lightning Casualties

The spatial locations of lightning casualty have been extracted through GIS mapping. The casualty rate has been fixed through no. of reporting from the particular

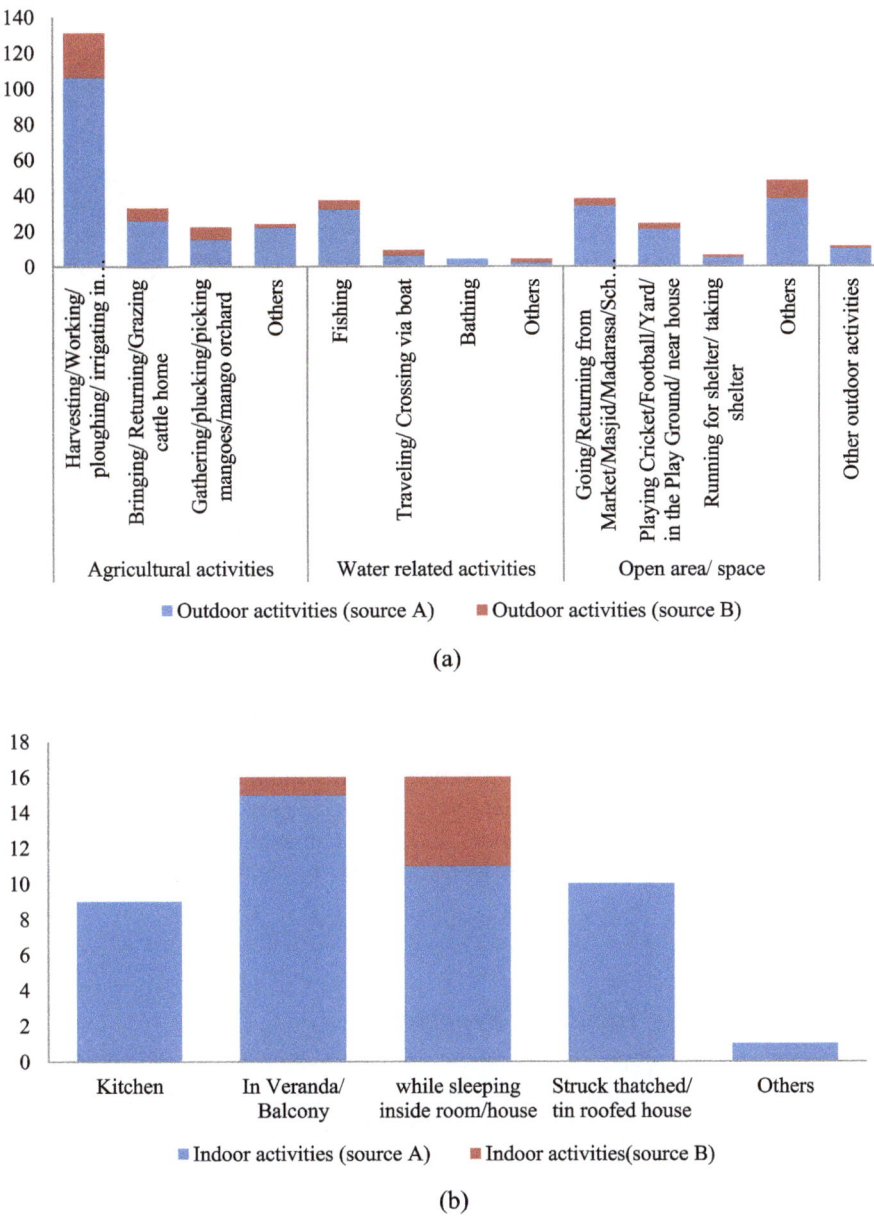

Fig. 10.8 Outdoor (**a**) and indoor (**b**) activities related to lighting casualties

area where lightning incidents occurred. The casualty rates have been divided into four categories: no casualties having zero casualty (green), low casualties (1–11) (light green), moderate casualties (12–21) (orange) and high casualties (22–40)

Fig. 10.9 Weather events reported during lightning incidents

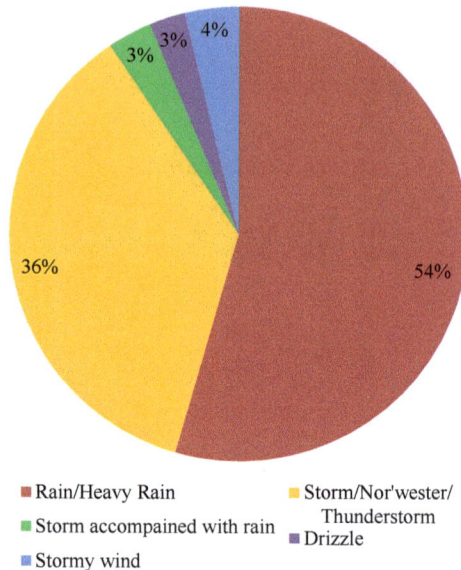

Rain/Heavy Rain   Storm/Nor'wester/Thunderstorm
Storm accompained with rain   Drizzle
Stormy wind

(red). District with the highest number of casualities are Chapai Nawabaganj, Kishoreganj and Sunjamganj, which means lightning reported from these areas are as high as 22–40 reports, moderate casualties in Thakurgaon, Dinajpur, Rangpur, Lalmonirhat, Gaibandha, Naogaon, Mymensingh, Habiganj, Maulavibazar, Brahmanbaria, Jhenaidah, Satkhira, Chandpur and Chittagong accounting 12–21 reports followed by rest other districts having low casualty reports except for districts like Feni, Rangamati, Shariatpur, Jhalokati, Patuakhali and Barguna had no casualties reported for lightning incidents, during the years of the study period (Fig. 10.10).

Annual Lightning Incidents Reported

Figure 10.11 shows the amount of lightning incident reported in the years 2011 and 2012 was high but the years 2013 and 2014 had comparatively lower lightning incidents reported compared to the previous two years. Lightning incident reporting drastically increased for the years 2015 and 2016 as per the study.

The deaths and injuries occurred due to lighting as shown in Fig. 10.1a, b show that the number of deaths due to lightning is higher than the injuries. But it is said that the numbers of injuries are as high as deaths but compared to fatalities the injuries are reported less (Mills et al. 2008). According to Singh and Singh (2015), there are more people injured by lightning strike than fatalities per year.

Figure 10.2a, b shows that the deaths and injuries among males are greater than that of females; this is most likely because larger proportions of males are involved

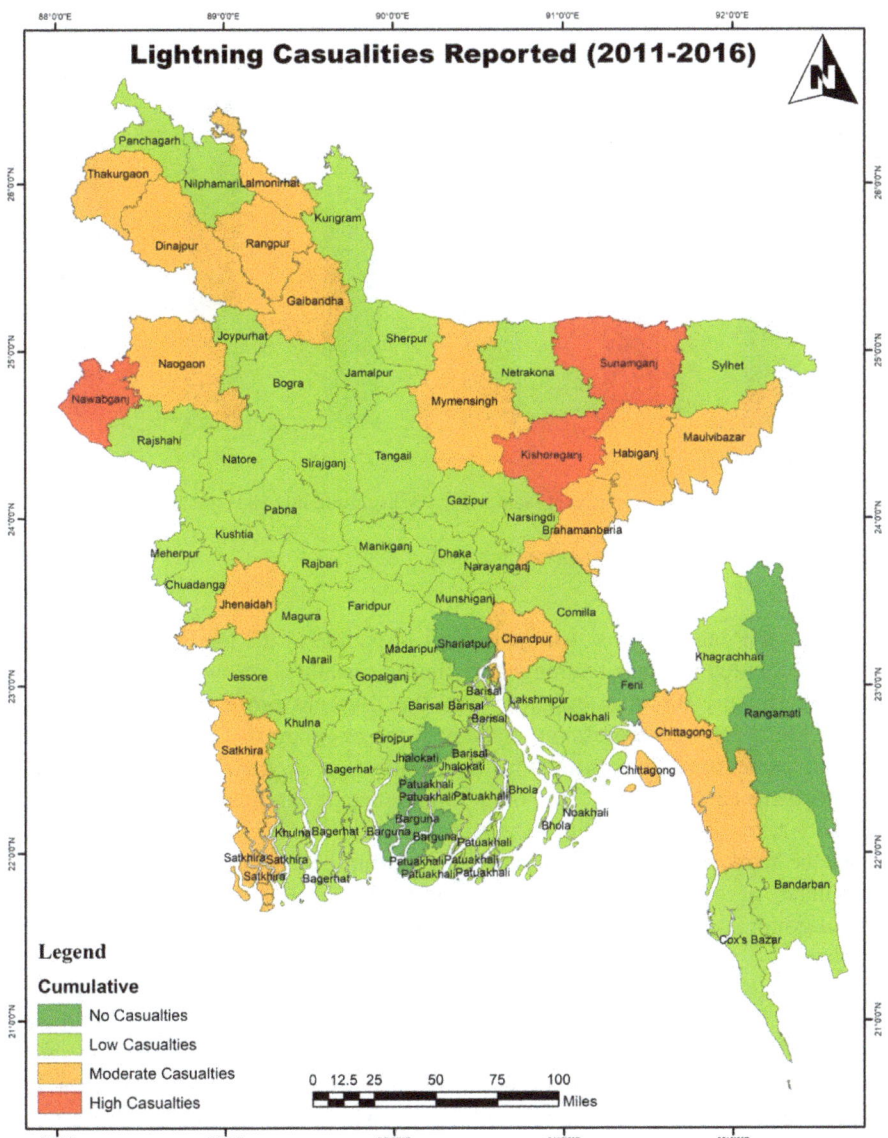

Fig. 10.10 Spatial distribution of lightining

with outdoor and traditional activities compared to females. The results are similar
to the findings of other regions where males accounted for larger amount of
casualties than female, e.g., in India (Singh and Singh 2015), Colombia
(Navarrete-Aldana et al. 2014), Canada (Mills et al. 2008), Swaziland (Dlamini
2009), UK (Elsom 2001), USA (Curran et al. 2000), etc.

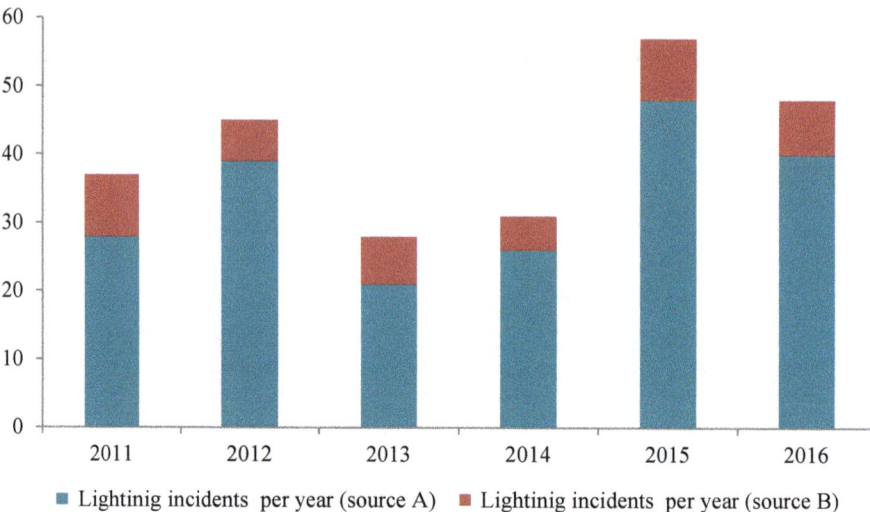

Fig. 10.11 Lightning incidents reported per year

During the data analysis, a large portion of people of the age groups 16–30 and 31–45 accounted for most amounts of lightning casualties (shown in Fig. 10.3); this particular feature has a co-relation with the different professions of the people (Fig. 10.7). From the profession analysis, a major group of people were from the farming profession, and during the data collection process majority number of farmers were aged from 20, 25, 30 and 35. Hence, it can be said that the lightning causality rate of people from the age groups of 16–30 and 31–45 is high as they fall under the working class of the society. Age group below <16 accounts for a good percentage of lightning casualties as people from this age category are mostly school-going students walking from home to school or vice versa or playing outdoors.

From Figs. 10.4 and 10.5, lightning incidents have been seen to occur most during the morning and afternoon. According to Yamane et al. (2010), in all other prior studies around the world, lightning casualties have been seen to occur during the daytime maximum due to vertical instability resulting from heating of the land surfaces and leading to lightning producing convection. The monthly variation of lightning casualties peaked in the pre-monsoon (March–May) and monsoon (June–September) season as shown in Fig. 10.6. This peak in the pre-monsoon season is due to the occurrence of severe local convective storms during this season (or known as nor'wester) accompanied with hail, gusts, rainfall and lightning (Yamane et al. 2010). According to Mannan and De (1995), the violent thunderstorms occur during March through May as since it is a transition phase. During this time, temperature and evaporation increase as sun becomes more overhead during March to May and as Bangladesh is full of water bodies moisture support increases during these months, thus helping in convective developments, forming giant cumulonimbus clouds and resulting in severe thunderstorm (Mannan and De 1995). Another reason for this peak could be because of the cropping season during mid-March to mid-July (Kharif-1) (Hasanuzzaman n.d.), thus resulting in increased human activities outdoors and hence increased casualties occurring during these months.

Lighting casualties mostly occurred outdoors as per the findings of this study compared to indoor; the activities involved are similar to those in the developing countries. According to Gomes and Kadir (2011), substantial amount of time is spent by people outdoors, who depend on labor-intensive agricultural practices. Figure 10.8a shows activities involving outdoor activities, showing the most amount of lightning casualty is related to harvesting/farming/plowing, etc., followed by cattle herding, grazing and walking them to shelters. A good amount of people was affected by mango plucking or gathering. This particular activity is something that is done out of recreation and can actively avoid during the thunderstorm period. Water-related activities account for lesser lightning casualty compared to agricultural activities but should be acknowledged since many people are involved with fishing be it professionally or for recreational purpose; but traveling by boat is an important means of transportation for most of the rural people, and hence to avoid the lightning-related casualty it is important to ensure safe traveling through boat. Figure 10.8b depicts the indoor lightning casualties, and it is visible from the figure that people who are affected by lightning are not residing in lightning-protected houses. It is not possible to directly quantify the lightning-protected houses provided by Bangladesh housing, but it can be concluded by saying that not many housings are lightning safe and that an unspecified section of the rural Bangladeshi dwelling is unprotected to lightning (Dewan et al. 2017).

Since the lightning-related deaths and injuries occur mostly during the pre-monsoon season due to the occurrence of severe thunderstorms during these periods of the season, the weather conditions recorded in Fig. 10.9 even though scattered stand true for the higher percentage of storms/thunderstorms and rainfall during lightning incidents.

According to Chowdry and Karmakaras (1986), as mentioned by Yamane et al. (2010) nor'wester storms occur during the pre-monsoon season of Bangladesh, most routinely in the north central region of Bangladesh. The southern region of Bangladesh experiences SW/S wind during the pre-monsoon season which with the help of sea moisture and other water bodies in that region gets enough moisture to build up thunderstorm activities. In the western side of Bangladesh, there lies Gangetic Plain of West Bengal, due to the presence of hilly tracts in Bihar and Orissa; thunderstorm develops due to the downdrafts of hilly region as it mixes with the sea moisture laden in the southeast surface winds. Nor'wester downdrafts are also experienced in the middle region of Bangladesh with the help of moisture form the Padma and Meghna rivers causing thunderstorm activities. The thunderstorm activities are also in the northern part of the Bangladesh due to the wind system present in the foothill region of Himalayas (Mannan and De 1995). These thunderstorm activities show how geographically Bangladesh falls under severe thunderstorm activities during the pre-monsoon season, so casualty form lighting is eminent; but this rate of casualty due to lightning can be reduced if awareness is built and structural solutions are given so that people can seek for shelter during these thunderstorm activities.

It cannot be said if lightning incidents have increased from the past years due to lack of data availability, and further analysis of other weather events for those years can actually determine if the weather events had any co-relation between the incidents reported for those years (specifically 2015 and 2016); but from Fig. 10.11, it can be said that the reporting has increased from previous years but there is still scope of underreporting due to limited geographical coverage as it is visible from the graphical data presented in my study.

Policy, Plan and Action

Bangladesh regulative framework provides relevant legislative, policy and best practice framework for disaster management under which the activity of Disaster Risk Reduction and Emergency Management in Bangladesh is managed and implemented. The regulative framework requires more articulation in addressing the risk associated with lightning. The National Disaster Management Policy (National Disaster Management Policy 2015) mentions about nor'wester and Lighting Hazard Management, but incorporating lightning with other natural hazard hinders the attention it should get as one of the main natural hazards in Bangladesh. The National Plan for Disaster Management ('National Plan for Disaster Management (2016–2020) Draft' n.d.) very recently added lightning as one of the main hazards of Bangladesh. Setting up specific research center to understand the 'lightning' phenomenon and integrating 'lightning forecasting and early warning system' will help achieve the priorities enlisted in the plan of the NDMP 2016–2020 (Draft), and it will also help in better understanding of the risk associated with lightning and also achieve government aims to protect the vulnerable people from the adverse effect of natural calamity and in building resilience of the poor and reduce their exposure and vulnerability to environmental shocks, geo-hydro-meteorological hazards, emerging hazards, man-made disasters and climate-related extreme events. The Disaster Management Act (National Disaster Management Act 2012) does not have lightning mentioned under its definition of disaster, and the act needs to be revised and updated with specific addition of 'lighting' mentioned under its definition of disaster so that lightning is acknowledged as a single disaster event and gets the adequate amount of attention and awareness it requires throughout the nation. In the Standing Orders on Disaster (SOD) (Standing Orders on Disaster 2010), roles and responsibilities of the ministries, committees and organizations are described. As per the report of Palma (2016), the Government of Bangladesh had declared lightning as one of the natural disasters; the report also mentions government compensation of 20,000 TK according to the standing orders to be provided to the families who lost a family member due to lightning. But compensating 20,000 TK would not be enough if the wage earner of the family is lost due to lightning. Hence, the SOD should have specification and categorization on compensating families depending on their socioeconomic status and specify the duty to the assigned ministry responsible for the allocation of such funds. If the lightning-affected families are not adequately compensated or provided with monetary resources, it will result in increased poverty. This will hinder the government vision of eradicating poverty which is fundamental to the vision of the Government of Bangladesh.

Conclusion

Lightning casualties in Bangladesh are related to accidental circumstances due to lack of awareness, but they are mainly influenced by the geographical location and activity of the people. A better perception and understanding of lightning-related casualties will help in generation of public policies, regulations, safety measures, construction of lightning-protected infrastructures, early warning systems and significant public awareness. Early warning system will help alert public on the availability of the cumulonimbus clouds (CB clouds/thunder clouds) in the sky forecasted for the day, which people can avoid by limiting their outdoor activities. In Bangladesh, the socioeconomic status puts its people under the threat of lightning. Bangladesh is a developing country, and its people are still involved with labor-intensive agricultural practices, which force them to go out for work even in the severe weather conditions. In this case, the government can put up lightning safe shelters so that the people can take shelter during such severe weather events and provide lightning-protected boats that will allow and ensure safe means of transportation when traveling through waterways. Lightning-related deaths and injuries have become an extensive disaster in Bangladesh. So, it is required to study the threat that lightning brings to the society and to recognize how it may become an important determinant of social and economic welfare which may result in even higher damages in the absence of adequate action time ahead.

References

Allaby M (2003) Floods (Facts on file dangerous weather series). Facts on File Science and Library, New York

Cardoso I et al (2014) Lightning casualty demographics in Brazil and their implications for safety rules. Atmos Res 135–136:374–379 (Elsevier B.V). https://doi.org/10.1016/j.atmosres.2012.12.006

Chao TC, Pakiam JE, Chia J (1981) A study of lightning deaths in Singapore. Singapore Med J 22 (3):150–157

Coates L, Blong R, Siciliano F (1993) Lightning fatalities in Australia, 1824-1991. Nat Hazards 8 (3):217–233. https://doi.org/10.1007/BF00690909

Curran EB, Holle RL, Lopez RE (2000) Lightning casualties and damages in the United States from 1959 to 1994. J Clim 13(19):3448–3464. https://doi.org/10.1175/1520-0442(2000)013%3c3448:LCADIT%3e2.0.CO;2

Dewan A et al (2017) Recent lightning-related fatalities and injuries in Bangladesh. Weather Clim Soc 9(3):575–589. https://doi.org/10.1175/WCAS-D-16-0128.1

Dlamini WM (2009) Lightning fatalities in Swaziland: 2000-2007. Nat Hazards 50(1):179–191. https://doi.org/10.1007/s11069-008-9331-6

Elsom DM (2001) Deaths and injuries caused by lightning in the United Kingdom: analyses of two databases. Atmos Res 56(1–4):325–334. https://doi.org/10.1016/S0169-8095(00)00083-1

Gomes C, Kadir MZAA (2011) 'A theoretical approach to estimate the annual lightning hazards on human beings. Atmos Res 101(3):719–725 (Elsevier B.V.). https://doi.org/10.1016/j.atmosres.2011.04.020

Gomes C, Hussain MAF, Abeysinghe KR (2006) Lightning accidents and awareness in South Asia: Experience in Sri Lanka and Bangladesh. In: Proc 28th intl conf on lightning protection (September 2006), pp 1240–1243. https://doi.org/10.1016/j.atmosres.2011.04.020

Holle RL (2016a) Lightning-caused deaths and injuries related to agriculture. In: 24th international lightning detection conference and sixth international lightning meteorology conference, Vaisala, San Diego, pp 1–5

Holle RL (2016b) The number of documented global lightning fatalities. In: 24th international lightning detection conference and sixth international lightning meteorology conference. Vaisala, San Diego, pp 1–4

Holle RL (2017) Lightning fatalities in Bangladesh in may 2016. In: 2017 American meteorological society annual meeting, 8th conference on the meteorological applications of lightning data, Seattle, Washington, pp 10–13

Illiyas FT et al (2014) Lightning risk in India—challenges in disaster compensation. Econ Polit Wkly 23–27

Kithil R (1995) Lightning's social and economic costs—National Lightning Safety Institute. In: International aerospace and ground conference on lightning and static electricity, 26–28 Sept 1995. National Lightning Safety Institute. Available at: http://lightningsafety.com/nlsi_lls/sec.html

Mannan MAC, De UK (1995) Pre-monsoon thunderstorm activity over Bangladesh.pdf, vol 6, pp 591–606

Mills B et al (2008) Assessment of lightning-related fatality and injury risk in Canada. Nat Hazards 47(2):157–183. https://doi.org/10.1007/s11069-007-9204-4

National Disaster Management Policy (2015)

National Disaster Mangement Act (2012)

National Plan for Disaster Management (2016–2020) Draft (no date)

Navarrete-Aldana N, Cooper MA, Holle RL (2014) Lightning fatalities in Colombia from 2000 to 2009. In: 2014 international conference on lightning protection, ICLP 2014, pp 40–46. https://doi.org/10.1109/iclp.2014.6971992

Palma P (2016) Lethal lightning. The Daily Star, 30 Sept. Available at: http://www.thedailystar.net/frontpage/lethal-lightning-1291849

Raga GB, de la Parra MG, Kucienska B (2014) Deaths by lightning in Mexico (1979–2011): threat or vulnerability? Weather Clim Soc 6(4):434–444. https://doi.org/10.1175/WCAS-D-13-00049.1

Sabur AKMA (2012) Disaster management system in Bangladesh: an overview. India Q 68(1):29–47. https://doi.org/10.1177/097492841106800103

Salerno J et al (2012) Risk of injury and death from lightning in Northern Malawi. Nat Hazards 62(3):853–862. https://doi.org/10.1007/s11069-012-0113-9

Selvi S, Rajapandian S (2016) Analysis of lightning hazards in India. Int J Disaster Risk Reduction 19:22–24. https://doi.org/10.1016/j.ijdrr.2016.08.021

Singh O, Singh J (2015) Lightning fatalities over India: 1979-2011. Meteorol Appl 22(4):770–778. https://doi.org/10.1002/met.1520

Standing Orders on Disaster (2010) Ministry of Food and Disaster Management Disaster Management & Relief Division Disaster Management Bureau. https://doi.org/10.1097/00006199-195908020-00056

Tilev-Tanriover et al (2015) Lightning fatalities and injuries in Turkey. Nat Hazards Earth Syst Sci 15(8):1881–1888. https://doi.org/10.5194/nhess-15-1881-2015

Uman MA (2008) The art and science of lightning protection. Cambridge University Press

Walsh KM et al (2000) National Athletic Trainers' Association position statement: lightning safety for athletics and recreation. J Athletic Trainning 471–477

Gomes C, Kithil R (no date) Developing a lightning awareness program model third world based on American- South Asian experience

Hasanuzzaman M (no date) Cropping seasons

Yamane Y et al (2010) Severe local convective storms in Bangladesh: part I. Climatology. Atmos Res 95(4):400–406 (Elsevier B.V.). https://doi.org/10.1016/j.atmosres.2009.11.004

Chapter 11
Climate Change Impact and the Conservation of Marine Turtles: A Case Study from Teknaf, Bangladesh

Methila Sarker, Alifa Bintha Haque, Md. Nazrul Islam, and Md. Jakariya

Abstract The study was conducted to determine the perception of locals and fishermen toward marine turtle in Cox's Bazar district, Bangladesh. A revised map was created by mapping the location of important nesting sites according to local's given information. This study confirmed seven important nesting sites in Cox's Bazar district, which were identified using key informant interviews. Five species of sea turtles have been making nests in Bangladesh. The olive ridley (39%) was seen the most while green turtles, Hawksbill, and Loggerhead were also observed. A total of fifty-two people were selected randomly to be interviewed in order to assess their perception toward marine turtle. About 58% of the respondents thought that turtles were helpful while 4% of the respondents thought turtles were bad. The rest of the respondents (21%) answered that they were not sure whether sea turtle was beneficial or harmful while 17% said they are of no use. 88% respondents have a positive attitude toward turtle conservation while 12% respondents were oblivious. Exploring local perceptions has revealed that a high proportion of locals are aware of turtle's protection by law even though its endangered status is not well known subject. Results showed that all of fishermen did not catch turtles intentionally. Most turtles are entangled in current jal (38%) and basha jal (23%) accidentally. Basically, 43% of fishermen said turtles are important while the remaining 57% said they have no idea about its importance.

Keywords Climate change · Marine turtles · Coastal fisheries · Conservation · Bangladesh

M. Sarker · Md. Jakariya (✉)
Department of Environmental Science and Management, North South University, Bashundhara, Dhaka, Bangladesh
e-mail: md.jakariya@northsouth.edu

A. B. Haque
Department of Zoology, Faculty of Biological Science, Dhaka, Bangladesh

Md. N. Islam
Department of Geography and Environment, Jahangirnagar University, Savar, Dhaka 1342, Bangladesh

© Springer Nature Switzerland AG 2021   205
Md. Jakariya and Md. N. Islam (eds.), *Climate Change in Bangladesh*,
Springer Climate, https://doi.org/10.1007/978-3-030-75825-7_11

Background

Marine environments and their biodiversity are under increasing pressures and threats due to environmental change caused by anthropogenic and natural changes. At the time of Columbus' voyages to the Caribbean, sea turtles were so abundant that vessels that had lost their way could follow the noise of sea turtles swimming along their migration route and find their way to the Cayman Islands (King 1982). During a meeting of the IUCN MTSG in 2003, climate change was ranked last out of 12 "burning issues" related to sea turtle research and bycatch evaluation and mitigation was the top "burning issue" in 2003 (Mast et al. 2004); but in 2009, Marine Turtle Specialist Group through its burning issues assessment identified global warming as one of the top five major hazards to marine turtles (Poloczanska et al. 2009). Climatic aspects such as temperature, extreme weather events, precipitation, ocean acidification and sea level rise have potential to affect marine turtle populations (Hawkes et al. 2009).

The following effects of climate change will have critical implications for marine turtles:

Effects of climate change	Implications on marine turtles
Severe storms (Hurricanes and tropical cyclones)	Increase beach erosion rates, endangering sea turtle nesting habitat
Hotter sand	Higher temperatures cause the sand to heat up and lead to a higher proportion of female to male hatchlings
Sea level rise	Result in a large loss of beach nesting habitat
Warming oceans	Result in changing ocean current potentially introducing sea turtles to new predators and harming the coral reefs which some of them rely on for survival

Oceana (n.d); NASA's Global Climate Change (2019)

With increasing temperatures, scientists predict warming climate may drive a significant threat to genetic diversity by creating more female than male hatchling (Sea Turtle Conservancy, n.d). Recent researches such as the higher population of female turtles in the northern Great Barrier Reef (Williams 2018) and also reporting 84% of female youngsters on an African island nation of Cape Verde (Paquette 2019) have been blamed on climate change. Study suggests that up to half of the current available nesting areas could be lost with predicted sea level rise (Witt et al. 2010). On the other hand, apart from climate change, humans have caused marine turtle populations to decline significantly all over the world. Initially, direct fishing for marine turtles was the main reason for population declines (NRC 1990; Oravetz 1999) a lot of the marine turtles get entangled in the fishing nets, drown, and die of suffocation. Today, other threats, including development and alteration of the coastal habitat, consumption of marine turtle eggs, pollution, and climate change are making the turtle populations vulnerable throughout the world (Comer and Nichols 2007). Five species of marine turtles are found in the territorial water of

Bangladesh, namely olive ridley turtle (Lepidochelys olivacea), green turtle (Chelonia mydas), hawksbill turtle (Eretmochelys imbricata), loggerhead turtle (Caretta caretta), and leatherback turtle (Dermochelys imbricata) (Sarker and Hossain 1995) and all are included on the IUCN Red List of Threatened Animals (Baillie and Groombridge 1996). Sea turtles occupy a special niche in the marine ecosystem and have survived millions of years in this environment. Sea turtles clearly play important roles in marine ecosystems. Each sea turtle species uniquely affects the diversity, habitat and functionality of its environment. Whether by grazing on sea grass, controlling sponge distribution, feasting on jellyfish, transporting nutrients or supporting other marine life, sea turtles play vital roles in maintaining the health of the oceans. (Wilson et al., n.d.). The resulting population declines have reduced the species ability to fulfill their roles in maintaining healthy marine ecosystems.

In Bangladesh, fisheries contribute about nearly 3% to GDP and more than 8% to the export earnings of the country (Bangladesh Population and Housing Census 2011, 2015). Marine fish contributes about 20% of total fish production in Bangladesh (Islam et al. 2001). Marine fisheries constitute of industrial fishery by large trawlers and artisanal fisheries by mechanized and non-mechanized boats (Barua et al. 2014). Incidental catch in fisheries is widely recognized as a major mortality factor for sea turtles with several gear types, including shrimp trawl nets and fish seines. (Oravetz). A large number of people in Bangladesh's coastal zone are involved in marine fishery. This is also causing an even larger number of bycatches being entangled (Phillott et al., n.d) in nets. This poses a huge threat to marine turtles that become trapped in these nets, as they face greater risks of being strangled or even drowned. Adding to the equation all previously mentioned threats, sea turtle populations are unfortunately heavily threatened worldwide. Despite our knowledge of the threats they face on a large scale, little is known about the extent of human impacts on sea turtles on a small scale, especially in tropical regions and the reasons behind their declining population.

Sea turtles were added to the protected list of the Bangladesh Wildlife (Preservation and Protection) Act in 2012 and should not be hunted or deliberately killed (Islam 2002a, b).

Also, Bangladesh is a signatory of the Convention on International Trade in Endangered Species of Wild Fauna and Flora (CITES), but still the turtles and tortoises are still indiscriminately being killed. Furthermore, Bangladesh recently signed the 'The Memorandum of Understanding (MoU)' for the Conservation and Management of marine turtles and their habitats of the Indian ocean and southeast Asia, but the Bangladesh Government is not yet meeting its commitments under this MoU (Islam 2001).

For years, dedicated researchers have conducted studies on sea turtles in and around Coxsbazar, which have proved to be worthwhile. Mohammad Zahirul Islam is among one of those researchers, who has done excellent research to evaluate threats to sea turtles in St. Martin's Island, trade in marine turtle products in Bangladesh, as well as on the nesting sea turtles at Sonadia Island. The latest paper concerning the present status of conservation and management of sea turtle in Co's

Bazar district, Bangladesh, was founded and published by M.A Hossain and others in 2013. The research was conducted to perceive the levels of knowledge and awareness about turtles, within the local communities and also to identify management shortfall and threats for sea turtle conservation and management in Cox's Bazar district. However, there is lack of data concerning revised information about the local's perception regarding marine turtles. We do not know whether they are aware of the increase or decrease in marine turtle population, why this issue, who are to be blamed, understanding of the laws and whether or not they are willing to work for the conservation. Furthermore, there have not been any recent studies reported, regarding current whereabouts or clearly mapped nesting sites of the marine turtles. On the other hand, Phillott et al have conducted extensive research to gather estimation of turtle bycatches in fisheries of Chittagong division, Bangladesh. However, there were no reports on the perception of fishermen. No knowledge of whether they knew about the importance of turtles or how and why they were dying. Through the perceptions of locals and fishermen, it is highly crucial to identify the main stakeholders for turtles, for instance: who is blaming who, who is being benefitted or facing losses, who is actually responsible for killing turtles and what measures can be taken to prevent such incidents. In order to implement successful conservation; it is important to communicate with the local communities, as community-based conservations are vital for long term successes. Also, in fruition, governmental organizations and regional NGO's will be able to make use of the results from identifying and documenting the following in their future plans and efforts to improve conservation management in Coxsbazar district. It is also vital to study whether marine turtles can adapt to the rapid predicted change in climate in the coming century. A research on climate variability and their impacts on the life stages of marine turtles from past to present should be executed to get an idea of the adaptive measures to be taken for future climatic change. Study in Bangladesh is required to investigate the effect of local climatic conditions on the hatchling output of marine turtles and key habitats upon which turtles depend to acquire biodiversity conservation strategies for potential results.

Objective

General Objective

The general objective of the study is to assess the perceptions of locals and fishermen toward marine turtle in Coxsbazar and Teknaf.

Specific Objectives

Through my study, I am aiming to contribute relevant and revised information about:

- Important sea turtle nesting sites in Teknaf and Coxsbazar.
- Access local and fishermen perceptions toward marine turtles.
- Access impacts of rapid climate change coupled with high anthropogenic impacts on marine turtles.
- To formulate recommendations in better protecting and conserving sea turtles.

Hypothesis to Be Tested

In regards to local perception

- There is an association with education level and knowledge on importance of marine turtles.
- There is no association with gender and will to work for conservation of marine turtles.

  In regards to fishermen

- There is an association with net/gear used in fishing and turtle entanglement in net.

Overview of Chapters

This research paper is divided into six chapters viz. (i) Introduction, (ii) review of the literature, (iii) methodology, (iv) results and discussion, (v) conclusion, and (vi) recommendations. The next chapter provides a critical review of previously published work relevant to the study. The methodology chapter thoroughly describes the methods applied in the study. It covers the description of the study area, sampling technique, methods for data collection and techniques applied in the analysis quantitative analysis of data. In the results and discussions chapter, findings from the study are presented using illustrative items. It also attempts to determine the significance of the results and compare them with the findings of previous works. The conclusion chapter provides a summary of all of the understandings drawn out from the study. Lastly, recommendations are projected to benefit conservation management plans in order to encounter the problems regarding climate change impacts on marine turtles.

Introduction on Marine Turtles

Marine turtles are an important component of global biodiversity (CITES 1973). They are one of the incredibly few species on earth to have lived relatively unchanged through mass extinctions with some fossil records dating back to 230 million years ago similar to species that still persist today (Spotila 2004). Sea turtles are globally widespread and have varying uses, roles and relationships in different coastal communities around the world (Lück 2008). From being a main income and food source (Garland and Carthy 2010; Parsons 2000) to having ancestral and cultural significance (Rudrud 2010; Morgan 2007), marine turtles are experienced and inhere a range of interpretations by the people who interact with them. There are seven species of sea turtle which includes the loggerhead (Caretta caretta), green (Cheloniamydas), hawksbill (Eretmochelys imbricata), Kemp's ridley (Lepidochelys kempii), olive ridley (Lepidochelys olivacea), flatback (Natator depressus), and leatherback (Dermochelys coriacea) turtles. An eighth species, the black turtle or East Pacific greenturtle (Chelonia agassizii), is recognized by some biologists, but morphological, biochemical and genetic data published to date are conflicting, and the black turtle is currently treated as belonging to Cheloniamydas (Meylan and Meylan 1999). All seven species of sea turtles are included on the IUCN Red List of Threatened Animals (Baillie and Groombridge 1996): Kemp'sridley and the hawksbill are considered critically endangered; loggerheads, green turtles, olive ridleys, and leatherbacks are listed as endangered; and flatbacks are considered vulnerable (IUCN 2010). Sea turtles spend their entire lives in marine or estuarine habitats. Only females will emerge from the sea, mainly to lay eggs on warm sandy beaches of tropical or subtropical latitudes (GrahamKordich 2003). Beach selection is influenced by a number of abiotic and biotic factors such as easy accessibility from the sea with a high beach platform as to not get flooded by tides or the water table (Mortimer 1982) in addition to relative low abundance of predators and competition from other species of sea turtles (Mortimer 1982; Blamires and Guinea 1998).

Geographic Distribution

Sea turtles live in almost every ocean basin throughout the world, nesting on tropical and subtropical beaches (Table 11.1).

Table 11.1 Geographic distribution of marine turtles

Species	Location	Habitat
Olive ridley	Tropical regions of the Atlantic, Indian, and Pacific Oceans (rarely to central California); nearly unknown around oceanic islands	Mostly coastal; traveling or resting in surface waters
Green turtle	Atlantic Ocean, Gulf of Mexico, along Argentine coast, Mediterranean Sea, and Indo-Pacific	Tropical and subtropical areas near continental coasts and around islands
Hawksbill	Throughout central Atlantic and Indo-Pacific regions; most tropical of all sea turtles	Near coral reefs and rocky outcroppings in shallow coastal areas
Loggerhead	Worldwide except Antarctica	Coastal tropical and subtropical; ventures into temperate waters, to boundaries of warm currents
Leatherback	Found in the Gulf of Alaska and south of the Bering Sea in the northeastern Pacific to Chile in the southeastern Pacific. In the Barents Sea Newfoundland and Labrador in the North Atlantic; around Argentina and South Africa in the South Atlantic. Throughout the Indian Ocean; and to Tasmania and New Zealand in the southwestern Pacific	Highly oceanic, approach coastal waters only during breeding season
Kemp's ridley	Adults usually occur in the Gulf of Mexico. Juveniles and immatures range between temperate and tropical coastal areas of the northwestern Atlantic Ocean. Occasionally young turtles reach northern European waters and as far south as the Moroccan coast	Shallow areas with sandy or muddy bottoms rich in crustaceans

Source Sea Turtles Habitat & Distribution; South Asia Most Dangerous for Sea Turtles, 2011

Importance of Marine Turtles

Sea turtles play a very special role in sustaining the productive longevity of our marine ecosystem and have survived millions of years in this environment. Because they migrate thousands of kilometers and take decades to mature, turtles serve as important indicators of the health of coastal and marine environments on both local and global scales (Frazier 1999). As unique components of complex ecosystems, sea turtles serve important roles in coastal and marine habitats by contributing to the health and maintenance of coral reefs, seagrass meadows, estuaries, and sandy beaches (Meylan and Meylan 1999). The sea turtles control the populations of the jelly fish which devour huge amount of fishes. Secondly, the turtles keep the aquatic environment clean by scavenging on the dead and rotten organic materials (Hossain et al. 2013). Thirdly, sea turtles improve their nesting beaches by supplying a concentrated source of high-quality nutrients (Bouchard and Bjorndal 2000). Unhatched sea turtles' eggs provide limited nutrients in dune ecosystems, such as nitrogen, phosphorus, and potassium which allow for the continued growth of vegetation and subsequent stabilization of beach dunes (Hannan et al. 2007). Plant growth not only helps to stabilize the shoreline but also provides food for a variety of plant eating animals and therefore can influence species distribution (Bouchard and Bjorndal 2000). Lastly, sea turtles provide food for other animals both on shore and at sea. By carrying around barnacles, algae, and other similar organisms known as Epibionts, sea turtles provide a food source for fish and shrimp (Schofield et al. 2006). Terrestrial animals such as ants, crabs, rats, raccoons, foxes, coyotes, feral cats, dogs, mongoose, and vultures—are known to dig up unhatched nests (Wilson et al., n.d). The eggs provide a nutrient-rich source of food for these predators which in turn redistribute nutrients among dunes through their feces (Hannan et al. 2007).

Uses of Marine Turtles

Humans and sea turtles have evidently played valuable roles in each other's existence throughout time, ever since people have settled the coasts and delved into the oceans. Coastal communities have relied upon sea turtles, primarily for their eggs, as it contains high amounts of protein and other products for countless generations and continue to do so, to this day, in many coastal areas (Meylan and Meylan 1999). Sea turtles have been used since time immemorial for food (oil and protein) and other commodities such as bone, leather, oil, and shell (Frazier 1999). For hundreds of years, sea turtles have been an integral part of the Pacific's culinary culture and history; turtle meat is a traditional food, their bones are used for making tools, and their shells are used for decorative or ceremonial purposes (Brikke 2009). Sea turtles have immeasurable worth as cultural assets. Diverse societies have traditionally held sea turtles as central elements in their respective customs and beliefs (Frazier 1999).

Global Threats to Marine Turtles

Sea turtles are disappearing right before this generation's eyes, said Whit Sheard, Pacific counsel, and senior adviser for Oceana (Five Sea Turtle Populations Are Endangered 2011). Sea turtles face many threats, both on land and at sea (Marcovaldi and Thome 1999). Threats to marine turtles around the world today are primarily anthropogenic. Tens of thousands of turtles die every year after being accidentally captured in active or abandoned fishing gear (Meylan and Meylan 1999). Entanglement in nets, fishing line, ropes, and other debris poses a huge threat to sea turtle (Sheavly and Register 2007). Other threats to turtle foraging areas include vessel groundings, certain fishing techniques (e.g., bottom trawling, dropping traps or anchoring blocks indiscriminately on living reef), near shore construction (e.g., piers, marinas), shoreline armoring (e.g., jetties, seawalls), careless snorkeling and diving (e.g., touching, collecting, trampling), reef walking (subsistence gleaning of shallow reef organisms, common throughout the insular Pacific) (Gibson and Smith 1999). Around the world, the survival of seven species of sea turtle is threatened by a variety of man-induced factors, including the direct and indirect harvest of adults and juveniles, threats to eggs and hatchlings, the degradation or loss of nesting habitat, and pollution of the seas (Comer and Nichols 2007). South Asia Most Dangerous for Sea Turtles (2011) study found that increasing climate change are also endangering these population.

Laws

Conventions, Agreements, and MOU Involving Sea Turtles (Mortimer 2002).

The following conventions deal directly with conservation of species including sea turtles:

(a) Convention on the Conservation of Migratory Species of Wild Animals (CMS);
(b) Convention on International Trade in Endangered Species of Flora and Fauna (CITES);
(c) Memorandum of Understanding on the Conservation and Management of Marine Turtles and their Habitats of the Indian Ocean and Southeast Asia (IOSEA) is an agreement under Article IV, Paragraph 4, of the CMS.

Bangladesh

The waters around India, Bangladesh, and Sri Lanka are home to the world's most endangered sea turtles (South Asia Most Dangerous for Sea Turtles 2011). Bangladesh (20° 34′–26° 38′ N and 88° 01′–92° 41′ E) located on the northern side of the Bay of Bengal in mainland of Asia. It is one of the resourceful countries with its wide range of marine and aquatic biodiversity. Bangladesh has a coastal area of 2.30 million ha and a coastline of 714 km along the Bay of Bengal, which supports a large artisanal and coastal fisheries (Kabir 2006). Bangladesh is one of the resourceful countries with its wide range of marine aquatic biodiversities. There are about 1093 marine aquatic organisms where 44.35% are finfish, 32.23% shellfish, 15.10% seaweeds, and only 8.32% are other organisms including shrimps (Kabir 2006). Five species of sea turtles are found in the territorial water of Bangladesh, namely olive ridley turtle (Lepidochelys olivacea), green turtle (Chelonia mydas), hawksbill turtle (Eretmochelys imbricata), loggerhead turtle

Table 11.2 Characteristic of dominant species in Bangladesh

Species	Color	Shell shape	Weight (kg s)	Length (feet)	Diet
Olive ridley	Olive green	Almost round shaped	45	2	Crabs, jellyfish, algae
Green turtle	Brown	Smooth oval shape	160	3–4	Sea Grass and Algae
Hawksbill	Dark, greenish brown	Narrow shell	46–70	3	Sponges
Loggerhead	Reddish brown	Heart shape shell	135	3–4	Whelks and Conch
Leatherback	Black to dark blue with white or pink spots	Shell is inside	250–700	6	Sea Grass and Algae

Source LALOË (2015), Sea Turtle Facts (n.d.)

(Caretta caretta), and leatherback turtle (Dermochelys imbricata) (Sarker and Hossain 1995).

Characteristic of dominant species in Bangladesh

Each species of turtle characteristics is different from each other but they share many behaviors, especially those involved in reproduction (Meylan and Meylan 1999) (Table 11.2).

Threats in Bangladesh

A large number of sea turtles in Bangladesh today get entangled in the fishing nets; many even drown and die of suffocation. This is due to the increasing human activities in harvesting the marine resources, particularly fishes. Moreover, as a result of the continuous development and alteration of coastal zones, climate change, pollution, and poaching of eggs, the sea turtle populations are being heavily affected throughout the world (Table 11.3).

Table 11.3 Threats in Bangladesh

In the sea	On nesting beach	Climate Change
Fisherman killed turtle if it entangled in net to save their valuable net (Islam 2002a, b)	Nesting females were occasionally killed by stray dogs on coastal island (Rashid and Islam 2006)	Habitat loss (Hossain et al. 2013) due to inundation of nest by sea level rise (Sarker 2009)
Pollution (Oil spill and other pollution) and disposal of solid domestic and machinery waste (Hossain et al. 2013)	Traditional utilization by local coastal communities (Shanker and Pilcher 2003)	Species loss (Hossain et al., 2013) due to more number of female hatchling because of high temperature (Shankar 2003)
	Physical alteration and development of infrastructure along the coast (Rashid and Islam 2006)	
	Beach lighting has also increased which disturbed nesting females and is disorients hatchlings (Hossain et al. 2013)	
	Poaching of eggs (Islam 2002a, b)	
	Turtles are consumed by a group of people as a source of protein and for its delicacy (Rao 1987)	

Source: Hossain et al. (2013), Islam (2002a, b), Rao (1987), Rashid and Islam (2006), Sarker (2009), Shanker and Pilcher (2003)

Lists of the International Agreements, Conventions, Treaties, and Protocols Signed, Accessed, Ratified by the Government of Bangladesh, Which Directly Or Indirectly Affects Marine Turtles

- Convention on the Control of Trans boundary Movements of Hazardous Wastes and their Disposal, Basel, 1989.
- Convention on the Continental Shelf, Geneva, 1958.
- Convention on Wetlands of International Importance especially as Waterfowl.

  Habitat, RAMSAR, 1971.

- Convention on International Trade in Endangered Species of Flora and Fauna,

  Washington, 1973.

- Convention on Biological Diversity, Rio de Janeiro, 1992
- Convention on the Conservation of Migratory Species of Wild Animals, 1979.
- International Convention for the Prevention of Pollution of the Sea by Oil, 1954.
- Indian Ocean—Southeast Asian Marine Turtle MoU.
- Bangladesh Wildlife (Preservation and Protection) Act in 2012.

Study Site

The study area was decided after a thorough literature review and interviews with a key informant. The key informant has been working in the International Marinelife Alliance as a research assistant for the past four years. He has informed me about the locations of their marine turtle hatcheries and where the turtles come to nest. From the literature review, we found that the major nesting sites of marine turtles were St. Martin's Island, Teknaf, Bordal, Sonadia Island, Kutubdia Island, Pechardwip, Inani Beach, Moheskhali Island, and Shahporirdwip (Hossain et al. 2013). Mozumdar et al. (2015) mentioned in his paper about the establishment of a total of five turtle hatcheries at Pechardwip, Khurermukh, Hazompara, Bodormokam, and Sonadia Eastpara sites. The deduction confirms that these places with hatcheries are the nesting sites of marine turtles; hence, I chose my site for locals from one of these places. Islam (2002a, b) observed that sea turtle mortality is presumed due to fisheries bycatch in coastal areas in the Chittagong Division (southwestern coast) of Bangladesh, including St. Martin's Island, Cox's Bazar-Teknaf Peninsula, and Sonadia Island. Therefore, as my study sites, I chose Cox's Bazar and Teknaf to access fishermen perception.

Location Area and Boundary

Coxsbazar is located at the fringe of the Bay of Bengal and is bounded on the north by Chittagong district, on the east by Bandarban district and Myanmar, on the south and west by the Bay of Bengal (District Statistics 2011 Cox's Bazar 2013). The total area of the district is 2491.85 km^2 (Bangladesh Population and Housing Census 2011, 2015) and lies between 20° 43' and 21° 56' north latitudes and between 91° 50' and 92° 23' east longitudes (District Statistics 2011 Cox's Bazar 2013). Teknaf Upazila is under Cox's Bazar district and the total area is 388.68 km^2 (Bangladesh Population and Housing Census 2011, 2015) located in between 20° 23' and 21° 09' north latitudes and in between 92°05' and 92° 23' east longitudes (Bangladesh Economic Zone Authority 2016). It is bounded by Ukhia Upazila on the north, the Bay of Bengal on the south, Arakan state of Myanmar on the east, the Bay of Bengal on the west.

Climate and Vegetation

The climate remains hot and humid with some seasons of temperate weather. Annual average temperature and rainfall varies from maximum 34.8 °C to minimum 16.1 °C (District Statistics 2011 Cox's Bazar 2013). A district feature of Cox's Bazar is clearly exhibited by the occurrence of mangrove vegetation and the coastal vegetation along the peninsula beach is represented by sand dune vegetation (Warrick and Ahmad 2012) (Fig. 11.1).

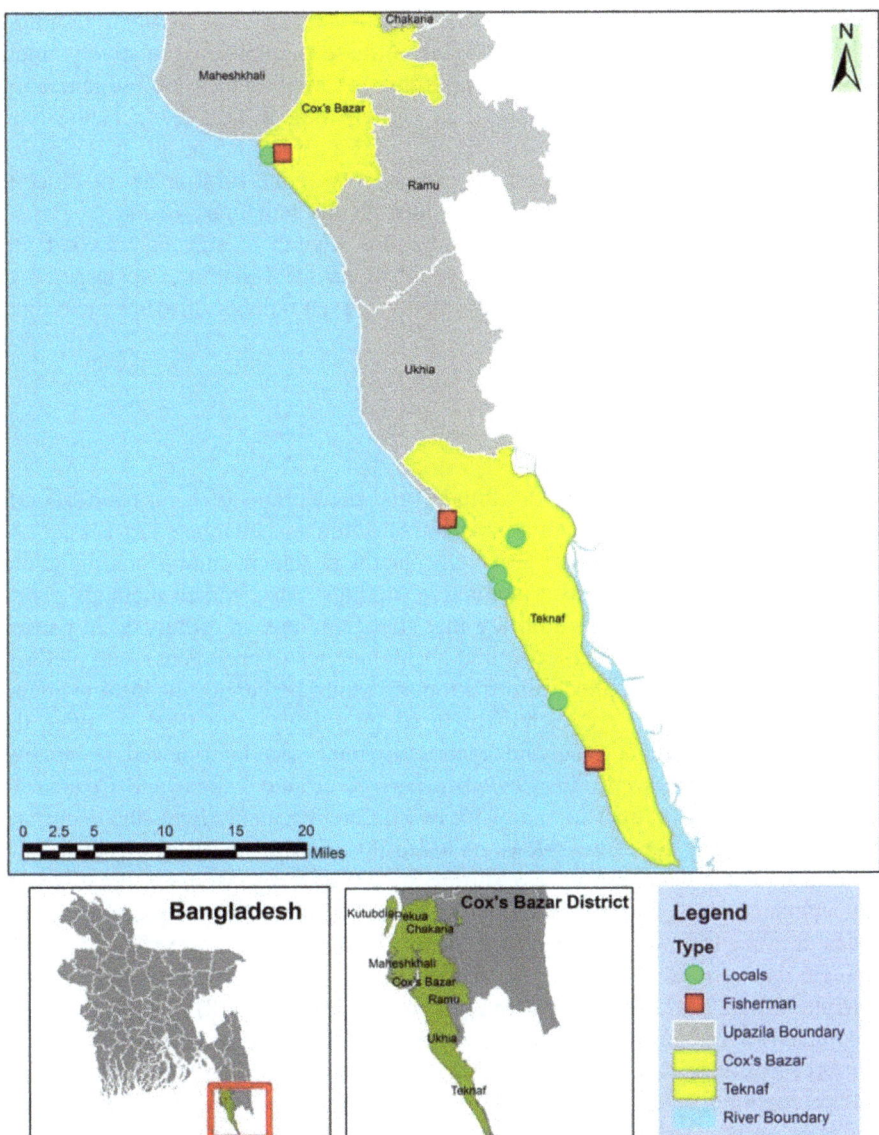

Fig. 11.1 Map of Cox's Bazar district showing the locations of the study sites

Sampling Technique

Based on the study site, the interviewees who were fishermen were selected by convenience sampling, i.e., persons encountered by chance at the place and time of the survey, and all the people approached accepted to be interviewed. Local residents were selected by using both the convenience and snowball sampling techniques. As the number of people living in these rural areas is relatively small, snowballing was identified as the most appropriate sampling measure for generating an appropriate sample size. But, prior to participating in this research, the local residents had to meet some basic criteria as well. Firstly, all the people participating had to be either temporary or permanent residents of the rural areas. In order to complete the survey, it was necessary to identify the temporary residents. Hence, two main criteria were taken into consideration: length of stay and second, the locals had to be 18 years or older. To obtain an ideal sample, we targeted to interview a variety of different age groups and an even number of women and men.

Data Collection

To gain a better understanding of attitudes and perceptions of local residents and fishermen, quantitative and qualitative data collection was conducted in Teknaf and Coxsbazar. Mixed methods are especially useful in understanding contradictions between quantitative results and qualitative findings. Also, mixed methods give a voice to study participants and ensure that study findings are grounded in participants' experiences. Two semi-structured questionnaires (Appendices) with pictorial guide were designed to collect information regarding perception of local residents and fishermen, respectively. Semi-structured interviews were used to make the respondent feel more at ease and each individual was interviewed separately. A total of 82 people were interviewed in person, around Teknaf and Coxsbazar. There were two sets of questionnaires for two different stakeholders—the locals and the fishermen. One set of interviews included 30 fishermen both from Teknaf and Coxsbazar. The main questions asked (directly or indirectly) were (1) What is your boat engine power? (2) How far do you go for fishing? (3) Do you see marine turtles in every trip? (4) Which type of fishing gear do you use? (5) Does your net entangle marine turtles? (6) How many dead and alive marine turtles you have seen in winter 2016? (7) Do you know the reasons of their dead? (8) Do you know the importance of marine turtles. Another set of interviews were conducted with 52 members (39 male and 13 female) of local communities. They were asked about their perception toward marine turtles. The main questions were (1) Which species you have seen? (2) Is the population of marine turtle increasing or decreasing? (3) Reasons of increase/decrease (4) When the number of marine turtles were high, what was the fate of the turtles? (5) Do you know the importance of turtles? (6) Do you want to save them? (7) What will you do to save them? (8) Is there any law regarding illegal poaching of marine turtles and their eggs?

Field surveys were carried out during April to September 2017 in Coxsbazar and Teknaf. Total I went three times for my field survey. Before conducting each interview, we explained the purpose of our study, guaranteed and assured the participant strict confidentiality and used only for academic purposes and only for the purposes of the particular research. To mitigate this potential bias, a reliable local citizen was chosen to remain present at all the interviews that were conducted. In doing so, we established a mutual feeling of trust and rapport, minimized any chance of miscommunication and acquired a sense of credibility. Information was collected from interviewee in a relaxed but guided manner. We followed the approximate order of survey questions but allowed participants to bring up topics as they wished. In general, interviews lasted approximately 15–20 min. The information was accumulated, grouped, and interpreted according to the objective as well as parameters. Some data contained numeric and some contained narrative facts.

Sources from secondary data are used to identify the primary impacts of the climate change on nesting beaches of marine turtles in Bangladesh and compare them to the primary data. Important marine turtle nesting grounds are identified through primary data collection to find out climate change affects in terms of distribution.

Mapping of Nesting Sites

Global positioning system (GPS) points of nesting site were taken at each site using a GPS device according to locals.

Statistical Analysis

Data was entered into Microsoft Excel. The collected data was then edited, summarized, and graphically analyzed with MS excel.

Chi-square test was used to assess whether the variables are associated using Statistical Package for the Social Science (SPSS).

In terms of the mapping data, ArcCatalog was used to convert the Microsoft Excel data into shape files which were then used in ArcMap to create GIS maps presented here.

Limitations

Financial: The research was self-funded so there was not sufficient money to go to field visit so often. Therefore, the sample size of the research was small as I could only visit the field thrice.

Access: There was difficulty in accessing the local people as majority of locals have relocated in Teknaf after the construction of marine drive. Also, it was difficult to access the documents of the organizations working for conservation of marine turtles in Coxsbazar district.

Socio-demographic Profile of Locals Profile of Locals and Fishermen

Of the 52 local respondents interviewed in six villages, the majority (75%) were males. Most of the respondents were youth (67%), whereas 19% were middle-aged and 14% were old. The majority of respondents were businessmen (55%) and the remainder includes homemakers (19%), fishing and related (17%), and others (12%). Others were employed as laborer, shopkeepers, farmer or teacher. The level of education of the respondents was low, (54%) of them having completed only elementary education mostly up to grade 5 or less and (34%) no education. 8% respondents were found to have secondary education and 4% Higher education. Overall, the coastal community was average monthly household income was Tk.10,000 or more. 13% income was less than 5k, 31% was 5–10K, 44% was 10–20K, and 13% was more than 20K. On average, the coastal community had five members in a family (Tables 11.4 and 11.5).

Table 11.4 Socio-demographic profile of locals

Explanatory variable				Villages				Total
		HP	JP	SP	PP	HC	SM	
Age	Youth (18–35)	5	7	3	4	3	13	35
	Middle-aged (36–50)	0	1	1	0	1	7	10
	Old (above 50)	2	2	0	1	0	2	7
Occupation	Fishing and related	0	1	1	3	0	4	9
	Business	1	4	1	0	1	9	16
	Homemakers	3	0	1	1	0	5	10
	Others	3	5	1	1	3	4	17
Sex	Male	4	10	3	3	4	15	39
	Female	3	0	1	2	0	7	13
Education level	No education	1	2	0	3	2	10	18
	Elementary	5	5	4	2	1	11	28
	Secondary	1	2	0	0	1	1	5
	Higher	0	1	0	0	0	0	1
Income level (Taka)	<5 K	0	0	1	1	0	3	5
	5–10 K	1	1	2	3	4	1	12
	10–20 K	2	4	0	0	0	11	17
	>20 K	1	2	0	0	0	2	5
No. of family members	≤ 3	0	1	2	2	1	5	11
	4–6	4	5	1	2	3	12	27
	≥ 6	3	4	1	1	0	5	14

Profile of the local respondents in six villages (HP = Hazampara, JP = Jahajpura, SP = Shaplapur, PP = Pochchim puranpara, HC = Hafizchora, SM = Somitipara)

Table 11.5 Socio-demographic profile of fishermen

Explanatory variable			Sites			Total
			Teknaf		Coxsbazar	
		Kururmukh	Mundhardhel	Shaplapur	Fisherighat	
Age	Youth (18–35)	3	9	1	6	19
	Middle-aged (3650)	0	1	1	7	9
	Old (above 50)	0	2	0	0	2
Experience (Years)	Novice ≤5	1	4	0	0	5
	Intermediate experience (6–10)	2	2	0	4	8
	Advanced experience ≥ 10	0	6	2	9	17

Of the 30 fishermen, 63% were youth, 30% middle-aged, and 7% were old. Most of the acknowledged were categorized as experienced, about 57% and belonged to Coxsbazar. The other 16% were novices, all belonging to Teknaf, whiles the rest of 27% were experienced fishermen.

Status of Marine Turtle Population According to Locals

Of those 30 locals, 72% of them said the turtle population is decreasing while the remaining 28% said they were increasing. However, the former group of people who claimed that marine turtle population is decreasing mostly have lower levels of education; 26% were not educated at all and the other 36% only having gone through elementary education. The people who claimed that marine turtle population is increasing, mostly have 16% elementary education level (Figs. 11.2 and 11.3).

Mostly, the reason of decrease according to the locals is fishing (59%). Incidental catch in fisheries is widely recognized as a major mortality factor for sea turtles. Several gear types, including shrimp trawl nets and fish seines, are known sources of injury and mortality (Oravetz 1999). 28% declination is due to anthropogenic activities. Coastal development, building construction, beach nourishment projects, erosion, and vehicular traffic are the main human-related activities impacting available nesting habitats.

People who claimed that the marine turtle population was increasing, was the minority, because they think that the eggs of turtles are in proper conservation. Henceforth, there are now NGOs that employ security capable enough to prevent poaching of the marine turtle eggs (Fig. 11.4 and Table 11.6).

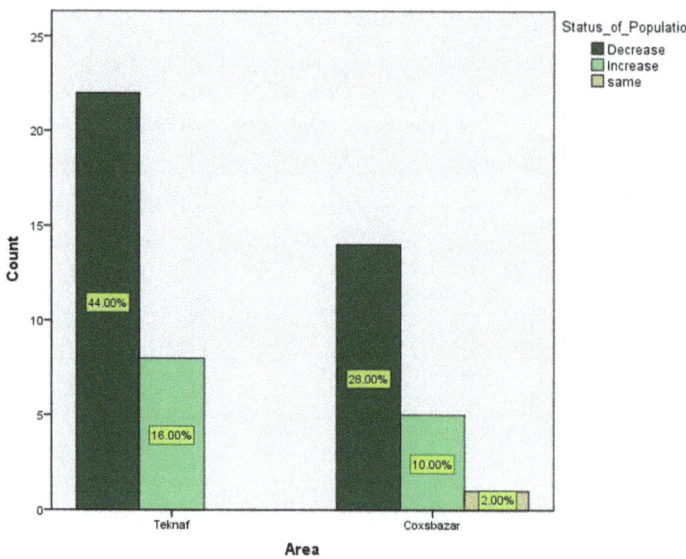

Fig. 11.2 Status of marine turtle population

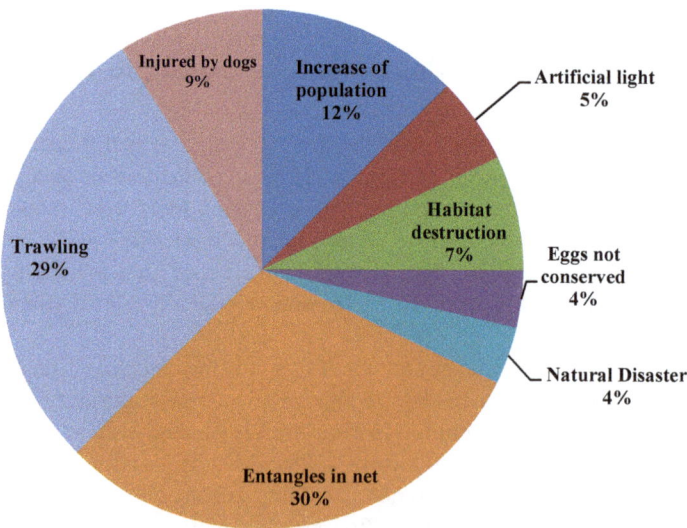

Fig. 11.3 Reasons of decrease in marine turtle population

The most common species is olive ridley (39%). Green turtle (22%) and Hawksbill (22%) too are regularly witnessed according to the locals. According to Islam (2002a, b), olive ridleys and green turtles were common while hawksbills

Fig. 11.4 Dominant species
according to locals

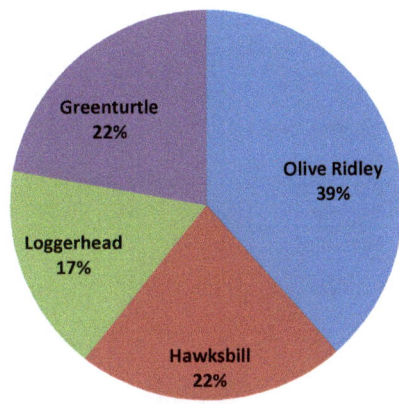

were rare. These nesting sites are important to find out the impacts of climate change on the marine turtles according to their distribution (Fig. 11.5).

62% said there were more marine turtles before 5–10 years as the population was lesser back then; beach development was slow and there were less fishing boats as well. This also shows that climate change is happening. Due to the ocean warming, ocean current are changing and marine turtles are changing their paths. 21% said that the no. of turtles were low before 5–10 years as turtle eggs are securely conserved nowadays. 13% have no idea whether turtle population is increasing or decreasing.

Fate of Marine Turtle and Their Eggs

85% said turtle eggs and meat are not sold in the market now because of strict laws. But 15% said that people still sell turtle and their eggs if they get the chance. The meat and eggs are consumed by Hindus, Maghs, and Chakmas (Fig. 11.6).

Knowledge of Locals on Importance of Turtles

The p-value (0.480) is more than 0.05 which indicates that these variables are independent of each other and that there is no statistically significant relationship between education level and knowledge on importance of marine turtles. Majority with no education and elementary level education claimed that they were helpful or good. Whereas (8%) with elementary qualifications and the other (8%) with no education said that they were of no use. Hence, this is where the confliction is noticed (Table 11.7).

Table 11.6 Nesting sites of dominant species

Species	Location
Olive Ridley	Shoibal point, Najiratek, Somitipara, Moheshkhali chor, Laboni beach, Baharchora, Ukhia, Inanai, Teknaf, Hazampara, Meyarpara, Jahajpura, Rajachora, Shaplapur, Hafizchara
Green turtle	Shonadia, Jahajpura, Hafizchora, Shaplapur
Loggerhead	Sabrang, Kururmukh
Hawksbill	Kutubdia, Somitipara, Hazampara, Jahajpura
Leatherback	Hazampara (dead)

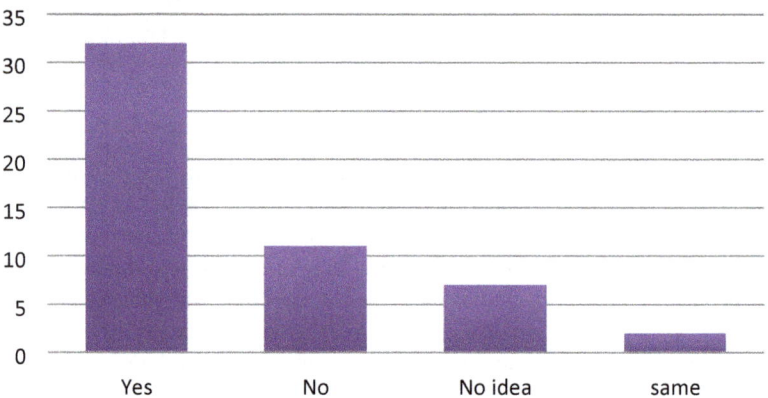

Fig. 11.5 Was there more turtles 5–10 years ago?

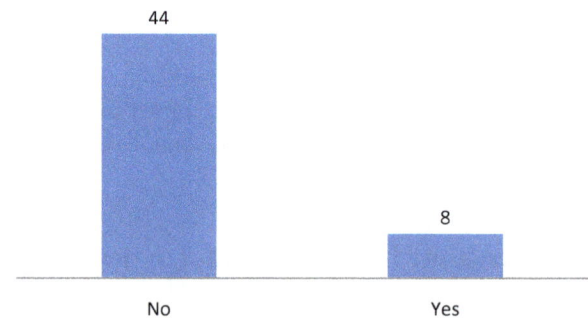

Fig. 11.6 Do people still sell turtle egg or meat in the market now?

About 57.7% of the respondent involved in business (21%) and other activities (19%) thought that turtles were helpful. Most of them replied that turtle ate waste, insect of the sea and kept clean the environment. 21% of local people had no clear idea about sea turtles and its importance. They answered that turtles are good creature of Allah and turtles added beauty of their area. About 4% of the respondent

Table 11.7 Knowledge of importance of turtles according to occupation

Occupation	Importance of turtles			
	Good (%)	Bad (%)	Helpful (%)	No use (%)
Fishing and related	3.8	1.9	9.6	1.9
Business	5.8	1.9	21.2	1.9
Homemakers	3.8		7.7	7.7
Others	7.7		19.2	5.8
	21.2	3.8	57.7	17.3

answered turtles were harmful because they ate fish and cut nets and created hazards. This group was constituted by fishermen and businessman. The rest of the respondents (17.3%) which included homemakers and others answered that sea turtle was of no use (Fig. 11.7).

The p-value (0.003) is less than 0.05 which indicates that these variables are not independent of each other and that there is statistically significant relationship between gender and will to save marine turtles. Males are more willing to work for conservation than females.

88% respondents said yes as they think turtles are good for the oceanic environment and do not harm them. 12% respondents said no because most of them really do not care as they do not know the importance.

Majority of the interested is knowledgeable enough to save turtles but people in general still need to be educated about the importance of turtles through workshops. Training needs to be provided to those who are unaware of how to protect turtles.

Fig. 11.7 Do you think marine turtles needs to be saved?

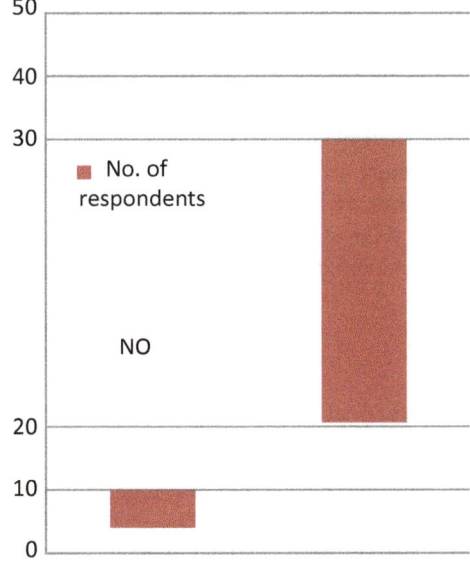

Fig. 11.8 Knowledge of
locals on laws regarding
marine turtles?

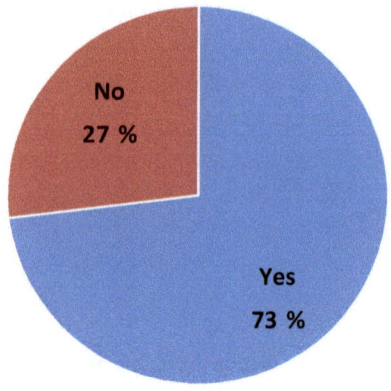

Strict laws against locals, especially fishermen killing turtles, should be regularly
monitored. And few said that preservation of the eggs is required (Fig. 11.8).

10% respondents said that people do not abide by the law because they are not
educated, environment laws are not taken seriously. People just show they are
abiding because there is no one to monitor regularly. Majority respondents (54%)
said that people abide by the law because people are scared, they will be punished.
36% said they do not know about any laws regarding marine turtles because it has
been only few years je laws are being publicize.

Knowledge of Fishermen on Importance of Turtles

43% said turtles are important because turtle eats waste and keeps the environment
clean. 57% said they do not know the importance. But majority answering "no"
thinks that turtle are harmful because they eat fish and tear off nets. Few indicated
turtle as a bad sign' because they think if they saw turtle during going to fishing
total amount of catch was significantly reduced. To get rid of these problem, he had
to arrange religious function (Muslim arrange "MILAD") (Figs. 11.9 and 11.10).

About 51% of the fishermen answered that turtles were caught in fishing net
while the rest of them (44%) answered that trawlers were killing turtles. Only 5%
said that predators are threat to the marine turtles and their eggs in the nesting
beach.

Discussion

St. Martin's Island on the southeast coast. Green turtle found nesting south-central
island beaches, mainland beaches in the southeast (from Cox's Bazar to Teknaf)
and also on some coastal islands like St. Martin's, Sonadia, Kutubdia, Hatiya, and

Fig. 11.9 Is there any importance of turtles

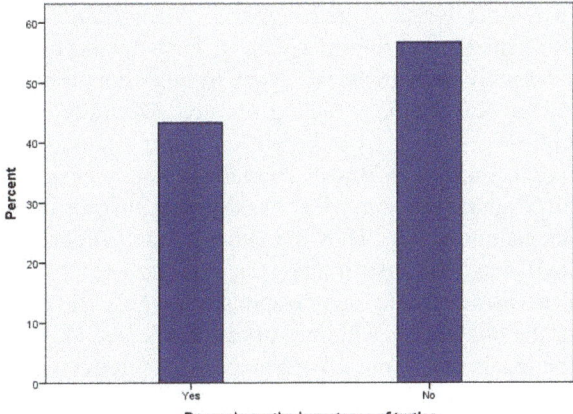

Fig. 11.10 Threats to marine turtles according to fishermen

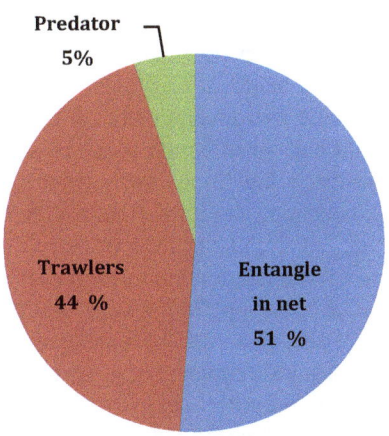

Sand weep islands. Many respondents (22%) said that they have seen hawksbill in Kutubdia, Somitipara, Hazampara, and Jahajpura. But the conservationist working for Marinelife Alliance did not mention about Hawksbill except for one person who said they can be seen at Coxsbazar but they just come around to visit, not to lay eggs. 62% locals said there were more turtles before 5–10 years as population was less back then, beach development was slow and there were not many fishing boats either. This data also points toward the impacts of climate change on nesting beaches. According to Hawkes et al. (2009), the aggravating threat in rise of sea level is a probability of an increase in fortification of coastal areas to protect human settlements which would reduce availability of sandy beaches to lay eggs. During a meeting of the IUCN in 2003, bycatch was the top "burning issue" of turtle decline (Mast et al. 2004); In fact, respondents (59%) said that is the reason for fishing declination. In 2000, the fisheries sector contributed about 6% to its national GDP (Bangladesh Population and Housing Census 2011, 2015). Hence, marine fishing is increasing because their demand is increasing. There are now more numbers of mechanized and non-mechanized boats to meet their demands. But this is not an issue only in Bangladesh. Even in other countries, the main reason for population declination is mostly fishing bycatch. Mortality of sea turtles entangled in Chilean gillnets is estimated to be 80% (Frazier and Montero 1990). In some parts of the world, such as in Brazil, coastal gill nets represent a larger mortality problem for turtles than trawling. Most of the turtles that get entangled in drift net and fixed gill net cannot escape. They die either due to suffocation or fishermen kill them to free their nets. Because if they tear off the net, 2–5 k is needed to repair it. Local fisherman normally deny responsibility for turtle deaths; they blame fishing trawlers or the big boats for killing turtles. But it was observed throughout the responses of the locals that about 20–22 boats turtles get entangled and they say that they let it go. But I am not sure if I can believe this statement of theirs. Sometimes, the superstitious fishing community considers the sighting of a turtle or a turtle getting entangled in the fishing nets as bad. Anthropogenic activities are the next after fishing to cause declination in marine turtle population. Now there is an increasing number of people and many progresses in coastal development, especially for tourists, beach nourishment projects, erosion, and vehicular traffic.

After the development of Teknaf's marine drive, there has been an increase in light and noise which is the reason for population decrease. Artificial lighting can misdirect hatchlings away from the sea and leave them vulnerable to dehydration, exhaustion, and predation. (Witherington 1999). It is seen even in other countries that development is the primary reason for turtle nesting issues. In Greece, extensive urbanization and resort development have forced turtles to compete with beachfront light, noise, vehicular traffic. Also, Turkey has many nesting beaches but they are threatened by increasing tourist development (Stachowitsch 2008).

Besides humans, it is observed that other predators are also responsible in destroying turtle's eggs such as dogs. This is not a new issue. During 1976–1982, Australia's exotic red foxes were responsible for devastating 90–95% loggerheads population on the east coast of Australia. (Stachowitsch 2008). Rashid and Islam (2006) identified that the threats to sea turtles in Bangladesh were stray dogs on nesting beaches, incidental capture of turtle, artificial lights, manmade physical alterations such as barriers around the beach, etc. Some people said that turtle population is increasing because eggs are being properly conserved. And in my observation, I have noticed that Marinelife Alliance and CREL are establishing hatcheries in many areas, for ex-situ breeding program, Marinelife Alliance. With the help of community, awareness is raised among the community people, local children's education through school programs, bycatch awareness program with the offshore fishermen through training and motivation. But many locals said that conservationists are stealing eggs and many eggs are destroyed in the hatchery. But then again, they said whenever they get informed about the discovery of a turtle or their egg, they buy it from the locals. So what I realized is, within these three stakeholders, there are a lot of perspective differences. Locals blame fishermen, fishermen blames trawlers and boats that are bigger that theirs and conservationists blame locals and fishermen.

Most of the people said that Maghs, Hindus ate turtle eggs. One of the respondent said shells are sold in Burma where they use it to make medicines. Interviewing a Hindu in Somitipara, said that they do not find it anymore so they do not have it. Many said turtle eggs are traded in Bandarban, Rangamati and Khagrachori, where it is consumed by maghs and chakmas. In this study, males were more interested about turtle management and conservation. Majority of local female said that they do not really care about turtle conservation. This can be because they do not go to the beach and less interaction with the turtles. There is no association of their education level or occupation with knowledge in importance of turtles. Because since 2–3 years, the project that was started by Marinelife Alliance was that project that gave a lot of people knowledge on the importance of turtles. Otherwise, I think long ago people knew a lot less about the benefits of turtles. Islam (2002a, b) found in his study that the nesting season runs from July until April. But majority locals and fishermen in my study said that turtles are mostly seen in winter. Last year 2016, the average number of turtles seen in winter was around 24. Respondents also said that lunar cycle is affiliated with turtle nesting. Turtles mostly come to lay eggs when it is full moon. But according to Ekanayake et al. (2002) and Law et al. (2010) there were no clear relationships between the number of nesting events and lunar phase. Majority use current jal (38%) and basha jal (23%). Fishermen were asked about the usage of Turtle Excluder Devices (TEDs) to avoid incidental capture of turtle which could reduce turtle mortality rate. Most of the respondent answered that they did not know anything about TED. TEDs were not used on any shrimp boats and the government was taking no initiative to impose any regulations regarding TED use. That is why the fisherman did not know much about it. We see that fishermen from Coxsbazar see turtles in every trip as they travel more than 100 km. Turtles are migratory animals traveling more than 1000 km just to lay eggs in their nesting habitat where they were born. Hence, the further you go, the higher the chances are of spotting turtles. Apart from everything, if we actually look into who are truly being benefitted from turtles, the answer is: conservationists, because they are getting paid to save the turtles and their eggs. Whereas fishermen are going through losses because most of the time turtles tear off their nets or if they use TEDs they might lose their targeted species. Locals are not being benefitted nor are they making any kind of profit. Yes, people of sell turtles, they are of course being benefitted. However, looking from an overall perspective, turtles are beneficial for all, seeing how its helps the environment. I have noticed a positive attitude in people toward turtle. If we can train them, educate them and find them an alternate income source then they will become more interested to work for turtle conservations. And most importantly, women can be used as a powerful tool in turtle conservation program.

Conclusion

The marine turtle populations of Bangladesh appear to be severely depleted due to a variety of threats, including direct exploitation for meat and eggs, habitat disturbance, and fishery bycatch. It is crucial to determine important nesting sites and assess threats to marine turtles in order to properly conserve sea turtle populations. Certainly, in Bangladesh, there are sufficient laws to protect marine turtles, although probably not sufficient to protect their habitats. Strict enforcement of protective laws and monitoring is required. But laws alone will not be effective; to have a successful conservation program it is essential to have community based conservation, which needs the understanding of perceptions of the local community. From a conservation point of view, the government should take initiatives for community based conservation and management including various awareness programs. Simply put, conserving sea turtles means protecting the seas and coastal areas, which in turn means protecting a complex, interconnected world in which human societies depend.

Recommendations

- Increase research efforts on the critical knowledge gaps about the threats from climate change which is influencing population numbers of the marine turtles. Current knowledge and status of sex ratios of marine turtles in Bangladesh and species distribution should be studied. We need to look for evidences how change in the ocean—atmosphere system associated with climate change, such as alternation of rainfall and storm patterns, sea level rise, and increase in ocean salinity impact the various life stages of marine turtles. These researches will provide critical information for the prediction of the potential impacts of climate change on marine turtles, which will help to form adaptive management practices.
- Based on results of threat assessment, it is necessary to develop appropriate programs in order to encourage community participation. To implement long term successes in conservation, it is essential to access threats via means of communication with the community. Observations, collected data and analyzed threats from previous reports should be taken into consideration which will then enable the development of an appropriate conservation program that will focus on sustainability.
- Minimize accidental mortality in fishing gear:

    TEDs and other "turtle friendly fishing technologies can be introduced to reduce bycatch. Fishermen are wary of using the device as they fear that their target catch will also be lost through the turtle escape hole. Hence, it is absolutely vital for fishermen to participate in reducing bycatch through gear modification or changes in their fishing procedures as this is a new thing to them. Furthermore, collaboration within all stakeholders is essential in reducing bycatch as well, since most exchanges of information primarily takes place in stakeholder workshops.

Bycatch Reduction Devices (BRDs) are essential for reducing the negative impacts of trawling on sustainability of marine resources and biodiversity. Several countries in the southeast Asia including Thailand, Philippines, Indonesia and Malaysia have been working toward the mandatory introduction of TEDs in shrimp trawlers to reduce turtle capture (Eayrs 2007). Use of BRDs in Bangladesh needs to be made mandatory in shrimp trawl nets and proper awareness generated in trawling industry about its necessity. Effective legislation and incentive schemes may be necessary for their popularization among fishermen.

- Reduce direct harvest of turtles and eggs and protect their habitats:

  Conduct public awareness and educational campaigns to enhance public awareness about turtle conservation in local communities, as well as among politicians and government personnel.

  Publicize the laws regarding marine turtle conservation to the local community

  Involve local communities in sea turtle surveys, monitoring, and management

- Develop financially self-sustaining sea turtle conservation and monitoring programs by creating jobs and introduce new environmentally friendly sources of income for communities near sea turtle habitats. In addition, develop activities to generate income and employment, significant to the presence of marine turtles, particularly through eco-tourism.
- Develop an umbrella organization at the national level to coordinate turtle conservation activities and to ensure multi-sectoral collaboration between government institutions, NGOs to protection of sea turtles and their habitats.
- Improve local legislation which currently has inconsistencies and gaps and inadequate fines. Regular monitoring of people abiding the law is a necessity.

References

Baillie J, Groombridge B (1996) IUCN red list of threatened animals. World Conservation Union (IUCN), Gland, Switzerland, 368 pp

Bangladesh Economic Zone Authority (2016, June) Social impact assessment for the Sabrang, Cox s Bazar economic zone

Bangladesh Population and Housing Census 2011 (2015) Dhaka: The Bangladesh Bureau of Statistics (BBS), Statistics and Informatics Division and Ministry of Planning

Barua S, Karim E, Humayun NM (2014) Present status and species composition of commercially important finfish in landed trawl catch from Bangladesh marine waters. Int J Pure Appl Zool 2 (2)

Bjorndal KA, Jackson JBC (2003) Roles of sea turtles in marine ecosystems: Reconstructing the past. In: Lutz PL, Musick JA, Wyneken J (eds) The biology of sea turtles, vol II. CRC Press, Boca Raton, Florida (USA), pp 259–273

Blamires SJ, Guinea ML (1998) Implications of nest site selection on egg predation at the sea turtle rookery at Fog Bay. In: Kennett R, Webb A, Duff G, Guinea M, Hill G (eds) Proceedings of the marine turtle conservation and management in Northern Australia Workshop. Centre for Indigenous and Natural Resources, Centre for Tropical Wetlands Management, Darwin, 20–24

Bouchard SS, Bjorndal KA (2000) Sea turtles as biological transporters of nutrients and energy from marine to terrestrial ecosystems. Ecology 81:2305–2313

Brikke S (2009) Local perceptions of sea turtles on Bora Bora and Maupiti islands, French Polynesia. SPC Tradit Mar Resour Manage Knowl Inf Bull 26:23–28

CITES (Convention on International Trade in Endangered Species of Wild Fauna and Flora) (1973) Text of the convention signed at Washington, D.C., on 3 March 1973 (Amended at Bonn, on 22 June 1979). http://www.cites.org/eng/disc/text.shtm

Comer KE, Nichols WJ (2007) Loreto bay: a refuge for the world's sea turtles. Loreto: the future of the first capital of the Californias, 47

District Statistics 2011 Cox's Bazar (2013) Dhaka: The Bangladesh Bureau of Statistics (BBS), Statistics and Informatics Division and Ministry of Planning

Eayrs S (2007) A guide to bycatch reduction in tropical shrimp-trawl fisheries. Food & Agriculture Org

Ekanayake EL, Ranawana KB, Kapurusinghe T, Premakumara MGC, Saman MM (2002) Marine turtle conservation in Rekawa turtle rookery in southern Sri Lanka. Ceylon J Sci (Biol Sci) 30:79–88

Five Sea Turtle Populations Are Endangered. Phys.org—News and Articles on Science and Technology, 16 Sept. 2011, phys.org/news/2011-09-sea-turtle-populationsendangered.html.\

Sea Turtle Facts (n.d.). Retrieved from http://seaturtleexploration.com/explore-and-learn/sea-turtle-facts/

Frazier JG (1999) Community-based conservation. Research and management techniques for the conservation of sea turtles. IUCN/SSC Mar Turtle Spec Group Publ 4:15–18

Garland KA, Carthy RR (2010) Changing taste preferences, market demands and traditions in Pearl Lagoon, Nicaragua: a community reliant on green turtles for income and nutrition. Conserv Soc 8:55–72

Gibson J, Smith G (1999) Reducing threats to foraging habitats. Research and management techniques for the conservation of sea turtles. IUCN/SSC Mar Turtle Spec Group Publ 4:184–188

Graham-Kordich K (2003) Local perceptions of the turtle conservation project in Gandoca-Manzanillo wildlife refuge, Costa Rica. University of Victoria, Victoria

Hamann M, Godfrey MH, Seminoff JA, Arthur K, Barata PCR, Bjorndal KA, Casale P (2010) Global research priorities for sea turtles: informing management and conservation in the 21st century. Endangered Species Res 11(3):245–269

Hannan LB, Roth JD, Ehrhart LM, Weishampel JF (2007) Dune vegetation fertilization by nesting sea turtles. Ecology 88(4):1053–1058

Hawkes LA, Broderick AC, Godfrey MH, Godley BJ (2009) Climate change and marine turtles. Endangered Species Res 7(2):137–154. https://www.researchgate.net/publication/222103019_Climate_Change_And_Marine_Turtles

Hossain MA, Miah MI, Hasan KR, Bornali JJ, Shahjahan M (2013) Present status of conservation and management of sea turtle in Cox's Bazar district of Bangladesh. Bangladesh J Anim Sci 42 (2):131–138

Islam MZ (2001) Notes on the trade in marine turtle products in Bangladesh. Mar Turtle Newslett 94(10)

Islam MZ (2002a) Marine turtle nesting at St. Martin's Island, Bangladesh. Mar Turtle Newslett 96:19–21

Islam MZ (2002b) Threats to sea turtles in St. Martin's island Bangladesh. Kachhapa 6:6–10

Islam MZ, Islam MS, Rashid SMA (1999) Marine turtle conservation program in St. Martin's island, Bangladesh by CARINAM: a brief review. Tigerpaper 26:17–28

Islam MS, Miah MTH, Haque MM (2001) Marketing system of marine fish in Bangladesh: an empirical study. Bangladesh J Agric Econ 24(1 & 2):127–142

Islam MZ, Ehsan F, Rahman MM (2011) Nesting sea turtles at Sonadia Island Bangladesh. Mar Turtle Newslett 130:19

IUCN (2010) IUCN red list of threatened species v. 2010.2. http://www.iucnredlist.org. Accessed 24 July 2010

Jackson JBC, Kirby MX, Berger WH, Bjorndal KA, Botsford LW, Bourque BJ, Bradbury RH, Cooke R, Erlandson J, Estes JA et al (2001) Historical overfishing and the recent collapse of coastal ecosystems. Science 293(5530):29–637

King FW (1982) Historical review of the decline of the green turtle and the hawksbill. In: Bjorndal KA (ed) Biology and conservation of sea turtles. Smithsonian Institution Press, Washington, D.C., pp 183–188

Laloë J (2015, July 1) Identifying sea turtle species. Turtles and Tides

Law A, Clovis T, Lalsingh GR, Downie JR (2010) The influence of lunar, tidal and nocturnal phases on the nesting activity of leatherbacks (Dermochelys coriacea) in Tobago, West Indies. Mar Turtle Newslett 127:12

León YM, Bjorndal KA (2002) Selective feeding in the hawksbill turtle, an important predator in coral reef ecosystems. Mar Ecol Prog Ser 245:249258

Lück M (2008) The encyclopaedia of tourism and recreation in marine environments. Centre for Agricultural Bioscience International and Credo Reference, Oxford, United Kingdom

Lynam CP, Gibbons MJ, Axelsen BE, Sparks CAJ, Coetzee J, Heywood BG, Brierley AS (2006) Jellyfish overtake fish in a heavily fished ecosystem. Curr Biol 16(13):R492–R493

Marcovaldi MAG, Thome JCA (1999) Reducing threats to turtles. In: Eckert KK, Bjorndal FA, Abreu-Grobois F, Donnelly M (eds) Research and management techniques for the conservation of sea turtles. IUCN/SSC Marine Turtle Specialist Group, Publication No.4, pp 165–168

Mast RB, Hutchinson BJ, Pilcher NJ (2004) IUCN/Species survival commission marine turtle specialist group news first quarter 2004. Mar Turtle Newsl 104:21–22

Meylan AB, Meylan PA (1999) Introduction to the evolution, life history, and biology of sea turtles. Research and management techniques for the conservation of sea turtles. IUCN/SSC Mar Turtle Spec Group Publ 4:3–5

Morgan CR (2007) Property of spirits: hereditary and globalvalue of sea turtles in Fiji. Hum Organ 66:60–68

Mortimer JA (2002) A strategy to conserve and manage the sea turtle resources of the Western Indian Ocean region. Report for IUCN, WWF and the Ocean Conservancy

Mozumdar PK, Islam MN, Mollah MA (2015) Conservation of sea turtle in Cox's Bazar-Teknaf Peninsula and Sonadia Island Ecologically Critical Area (ECA) of Bangladesh, 05th edn, vol 9, Ser. 2015). Retrieved from http://waset.org/pdf/books/?id=19016&pageNumber=136

NASA's Global Climate Change (2019, June 19). Climate change puts pressure on sea turtles. Retrieved from https://climate.nasa.gov/news/2879/climate-change-puts-pressure-on-sea-turtles/

National Research Council (1990) Decline of sea turtles: causes and prevention. National Academy Press, Washington, DC

Oceana. Sea turtles and climate change. Retrieved from https://usa.oceana.org/sea-turtles-and-climate-change

Oravetz CA (1999) Reducing incidental catch in fisheries. Research and management techniques for the conservation of sea turtles. IUCN/SSC Mar Turtle Spec Group Publ 4:189–193

Paquette D (2019, Oct 22) The warming climate is making baby sea turtles almost all girls. The Washington Post [online]. Retrieved from https://www.washingtonpost.com/world/africa/the-warming-climate-is-turning-baby-sea-turtles-one-gender/2019/10/21/d571f3fe-e3a6-11e9-b0a6-3d03721b85ef_story.html

Parsons J (2000) Sea turtles and their eggs. In: Kiple KF, Ornelas KC (eds) Cambridge world history of food, vol 1. Cambridge University Press, Cambridge, United Kingdom, pp 567–574

Phillott AD, Mathew JM, Krishnankutty N, Ara SS, Shathy ST, Akter T, Khan ZI (2015) Estimates of turtle bycatch in fisheries of Chittagong division, Bangladesh. Indian Ocean Turtle Newsl (22)

Poloczanska ES, Limpus CJ, Hays GC (2009) Vulnerability of marine turtles to climate change. Adv Mar Biol 56:151–211

Pritchard PCH (1980) The conservation of sea turtles: practices and problems. Am Zool 20 (3):609–617

Rajakaruna RS, Dissanayake DMNJ, Ekanayake EML, Ranawana KB (2009) Sea turtle conservation in Sri Lanka: assessment of knowledge, attitude and prevalence of consumptive use of turtle products among coastal communities. Indian Ocean Turtle Newsl 10:5–12

Rao RJ (1987) Ecological studies on Indian turtles. Tigerpaper 14(3):21–25

Rashid SMA, Islam MZ (2006) Status and conservation of marine turtles in Bangladesh. In: Shanker K, Choudhury BC (eds) Marine turtles of the Indian subcontinent. Universities Press, New Delhi, India, pp 200–216

Rudrud RW (2010) Forbidden sea turtles: traditional laws pertaining to sea turtle consumption in Polynesia (includingthe Polynesian outliers). Conserv Soc 8:84–97

Sarker MSU (2009) Climate change: impact on Bangladesh Published in daily newspaper in Bangladesh. http://www.thedailystar.net/newDesign/photo_gallery.php?pid=118260/Climate. change.impact.on.biodiversity.of.the.Sundarbans.htm. Accessed 15 Jan 2011

Sarker SU, Hossain ML (1995) Population, ecobiological status, captive propagation and conservation problems of Lissemys punctata in Bangladesh. International Congress of Chelonian Conservation. France (Proc), pp 43–46

Schofield G, Katselidis KA, Dimopoulos P, Pantis JD, Hays GC (2006) Behaviour analysis of the loggerhead sea turtle Caretta caretta from direct in-water observation. Endangered Species Res 2:71–79

Sea turtle conservancy. Information about sea turtles: Threats from climate change. Retrieved from https://conserveturtles.org/information-sea-turtles-threats-climate-change/

Shanker K, Pilcher NJ (2003) Marine turtle conservation in south and southeast Asia: hopeless cause or cause for Hope? Mar Turt Newsl 100:43–51

South Asia Most Dangerous for Sea Turtles: Study. Phys.org—News and Articles on Science and Technology, 29 Sept 2011, phys.org/news/2011-09-south-asia-dangeroussea-turtles.html

Sheavly SB, Register KM (2007) Marine debris & plastics: environmental concerns, sources, impacts and solutions. J Polym Environ 15(4):301–305

Spotila JR (2004) Sea turtles—a complete guide to their biology, behavior and conservation. The Johns Hopkins University Press, Baltimore and London

Stachowitsch M (2008) Sea turtles: a complete guide to their biology, behavior, and conservation. Mar Ecol 29(1):140–141

Welch C (2018, January 8) Rising temperatures cause sea turtles to turn female. National Geographic. Retrieved from https://news.nationalgeographic.com/2018/01/australia-green-sea-turtles-turning-female-climate-change-raine-island-sex-temperature/

Wildlife (Conservation and Security) ACT, 2012 (2012, July 10). Retrieved from http://extwprlegs1.fao.org/docs/pdf/bgd165019.pdf

Williams R (2018, Jan 10) Rising Temperatures And The Elimination Of Male Turtles [online]. The Scientist Magazine®. Retrieved from https://www.the-scientist.com/news-opinion/rising-temperatures-and-the-elimination-of-male-turtles-30433

Wilson, EG, Miller KL, Allison D, Magliocca M, WHYHEALTHYOCEANSNEEDSEATURTLES

Witherington BE (1999) Reducing threats to nesting habitat. Research and management techniques for the conservation of sea turtles. IUCN/SSC Mar Turtle Spec Group Publ 4:179–183

Witt MJ, Hawkes LA, Godfrey MH, Godley BJ, Broderick AC (2010) Predicting the impacts of climate change on a globally distributed species: the case of the loggerhead turtle. J Exp Biol 213(6):901–911

Wyneken J, Burke TJ, Salmon M, Pedersen DK (1988) Egg failure in natural and relocated sea turtle nests. J Herpetol 88–96

Lightning Source UK Ltd.
Milton Keynes UK
UKHW020641040822
406835UK00002B/15